Advanced Topics in Perpetual Motion

Contents

Chapter 1

Perpetual motion

For other uses, see Perpetual motion (disambiguation).

A **perpetual motion** is a motion of enduring physical ob-

Robert Fludd's 1618 "water screw" perpetual motion machine from a 1660 wood engraving. This device is widely credited as the first recorded attempt to describe such a device in order to produce useful work, that of driving millstones.[1] Although the machine would not work, the idea was that water from the top tank turns a water wheel (bottom-left), which drives a complicated series of gears and shafts that ultimately rotate the Archimedes' screw (bottom-center to top-right) to pump water to refill the tank. The rotary motion of the water wheel also drives two grinding wheels (bottom-right) and is shown as providing sufficient excess water to lubricate them.

jects that are very large compared with atoms, that continues indefinitely. This is impossible because of friction and other modes of degradation or disruption of form.[2][3][4] A

perpetual motion machine is a hypothetical machine that can do work indefinitely without an energy source. This kind of machine is impossible, as it would violate the first or second law of thermodynamics.[3][4][5]

These laws of thermodynamics apply even at very grand scales. For example, the motions and rotations of celestial bodies such as planets may appear perpetual, but are actually subject to many processes that degrade or disrupt their form, such as solar winds, interstellar medium resistance, gravitation, thermal radiation and electro-magnetic radiation.[6][7]

Thus, machines which extract energy from finite sources will not operate indefinitely, because they are driven by the energy stored in the source, which will eventually be exhausted. A common example is devices powered by ocean currents, whose energy is ultimately derived from the Sun, which itself will eventually burn out. Machines powered by more obscure sources have been proposed, but are subject to the same inescapable laws, and will eventually wind down.

1.1 History

Main article: History of perpetual motion machines

The history of perpetual motion machines dates back to the Middle Ages. For millennia, it was not clear whether perpetual motion devices were possible or not, but the development of modern theories of thermodynamics has shown that they are impossible. Despite this, many attempts have been made to construct such machines, continuing into modern times. Modern designers and proponents often use other terms, such as "overunity", to describe their inventions.

1.2 Basic principles

Main article: Thermodynamics

> Oh ye seekers after perpetual motion, how many vain chimeras have you pursued? Go and take your place with the alchemists.
> — Leonardo da Vinci, 1494[8][9]

There is a scientific consensus that perpetual motion in an isolated system violates either the first law of thermodynamics, the second law of thermodynamics, or both. The first law of thermodynamics is essentially a statement of conservation of energy. The second law can be phrased in several different ways, the most intuitive of which is that heat flows spontaneously from hotter to colder places; the most well known statement is that entropy tends to increase (see entropy production), or at the least stay the same; another statement is that no heat engine (an engine which produces work while moving heat from a high temperature to a low temperature) can be more efficient than a Carnot heat engine.

In other words:

1. In any isolated system, one cannot create new energy (first law of thermodynamics)

2. The output power of heat engines is always smaller than the input heating power. The rest of the energy is removed as heat at ambient temperature. The efficiency (this is the produced power divided by the input heating power) has a maximum, given by the Carnot efficiency. It is always lower than one.

3. The efficiency of real heat engines is even lower than the Carnot efficiency due to irreversible processes.

Statements 2 and 3 apply to heat engines. Other types of engines which convert e.g. mechanical into electromagnetic energy, could in principle operate with 100% efficiency, but only if they were to be free of friction. No such engine is possible in practice.

Machines which comply with both laws of thermodynamics by accessing energy from unconventional sources are sometimes referred to as perpetual motion machines, although they do not meet the standard criteria for the name. By way of example, clocks and other low-power machines, such as Cox's timepiece, have been designed to run on the differences in barometric pressure or temperature between night and day. These machines have a source of energy, albeit one which is not readily apparent so that they only seem to violate the laws of thermodynamics.

Machines which extract energy from seemingly perpetual sources - such as ocean currents - are indeed capable of moving "perpetually" until that energy source runs down. They are not considered to be perpetual motion machines because they are consuming energy from an external source and are not isolated systems.

1.2.1 Classification

One classification of perpetual motion machines refers to the particular law of thermodynamics the machines purport to violate:[10]

- A **perpetual motion machine of the first kind** produces work without the input of energy. It thus violates the first law of thermodynamics: the law of conservation of energy.

- A **perpetual motion machine of the second kind** is a machine which spontaneously converts thermal energy into mechanical work. When the thermal energy is equivalent to the work done, this does not violate the law of conservation of energy. However, it does violate the more subtle second law of thermodynamics (see also entropy). The signature of a perpetual motion machine of the second kind is that there is only one heat reservoir involved, which is being spontaneously cooled without involving a transfer of heat to a cooler reservoir. This conversion of heat into useful work, without any side effect, is impossible, according to the second law of thermodynamics.

- A **perpetual motion machine of the third kind**, usually (but not always)[11] defined as one that completely eliminates friction and other dissipative forces, to maintain motion forever (due to its mass inertia). *Third* in this case refers solely to the position in the above classification scheme, not the third law of thermodynamics. Although it is impossible to make such a machine,[12][13] as dissipation can never be 100% eliminated in a mechanical system, it is nevertheless possible to get very close to this ideal (see examples in the Low Friction section). Such a machine would not serve as a source of energy but would have utility as a perpetual energy storage device.

1.2.2 Impossibility

"Epistemic impossibility" describes things which absolutely cannot occur within our *current* formulation of the physical laws. This interpretation of the word "impossible" is what is intended in discussions of the impossibility of perpetual motion in a closed system.[14]

OCTOBER   25 CENTS

Popular Science MONTHLY

Perpetual Motion?
See page 26

October 1920 issue of Popular Science *magazine, on perpetual motion. Although scientists have established them to be impossible under the laws of physics, perpetual motion continues to capture the imagination of inventors. The device shown is a "mass leverage" device, where the spherical weights on the right have more leverage than those on the left, supposedly creating a perpetual rotation. However, there are a greater number of weights on the left, balancing the device.*

The principles of thermodynamics are so well established, both theoretically and experimentally, that proposals for perpetual motion machines are universally met with disbelief on the part of physicists. Any proposed perpetual motion design offers a potentially instructive challenge to physicists: one is certain that it cannot work, so one must explain *how* it fails to work. The difficulty (and the value) of such an exercise depends on the subtlety of the proposal; the best ones tend to arise from physicists' own thought experiments and often shed light upon certain aspects of physics. So, for example, the thought experiment of a Brownian ratchet as a perpetual motion machine was first discussed by Gabriel Lippmann in 1900 but it was not until 1912 that Marian Smoluchowski gave an adequate explanation for why it cannot work.[18] However, during that twelve-year period scientists did not believe that the machine was possible. They were merely unaware of the exact mechanism by which it would inevitably fail.

> The law that entropy always increases, holds, I think, the supreme position among the laws of Nature. If someone points out to you that your pet theory of the universe is in disagreement with Maxwell's equations — then so much the worse for Maxwell's equations. If it is found to be contradicted by observation — well, these experimentalists do bungle things sometimes. But if your theory is found to be against the second law of thermodynamics I can give you no hope; there is nothing for it but to collapse in deepest humiliation.
> — Sir Arthur Stanley Eddington, *The Nature of the Physical World* (1927)

The conservation laws are particularly robust from a mathematical perspective. Noether's theorem, which was proven mathematically in 1915, states that any conservation law can be derived from a corresponding continuous symmetry of the action of a physical system.[15] For example, if the true laws of physics remain invariant over time then the conservation of energy follows. On the other hand, if the conservation laws are invalid, then the foundations of physics would need to change.[16]

Scientific investigations as to whether the laws of physics are invariant over time use telescopes to examine the universe in the distant past to discover, to the limits of our measurements, whether ancient stars were identical to stars today. Combining different measurements such as spectroscopy, direct measurement of the speed of light in the past and similar measurements demonstrates that physics has remained substantially the same, if not identical, for all of observable time spanning billions of years.[17]

In the mid 19th-century Henry Dircks investigated the history of perpetual motion experiments, writing a vitriolic attack on those who continued to attempt what he believed to be impossible:

> "There is something lamentable, degrading, and almost insane in pursuing the visionary schemes of past ages with dogged determination, in paths of learning which have been investigated by superior minds, and with which such adventurous persons are totally unacquainted. The history of Perpetual Motion is a history of the fool-hardiness of either half-learned, or totally ignorant persons."[19]
> — Henry Dircks, *Perpetuum Mobile: Or, A History of the Search for Self-motive* (1861)

1.2.3 Techniques

See also: History of perpetual motion machines

Some common ideas recur repeatedly in perpetual motion machine designs. Many ideas that continue to appear today were stated as early as 1670 by John Wilkins, Bishop of Chester and an official of the Royal Society. He outlined three potential sources of power for a perpetual motion machine, "Chymical Extractions", "Magnetical Virtues" and "the Natural Affection of Gravity".[1]

The seemingly mysterious ability of magnets to influence motion at a distance without any apparent energy source has long appealed to inventors. One of the earliest examples of a magnetic motor was proposed by Wilkins and has been widely copied since: it consists of a ramp with a magnet at the top, which pulled a metal ball up the ramp. Near the magnet was a small hole that was supposed to allow the ball to drop under the ramp and return to the bottom, where a flap allowed it to return to the top again. The device simply could not work. Faced with this problem, more modern versions typically use a series of ramps and magnets, positioned so the ball is to be handed off from one magnet to another as it moves. The problem remains the same.

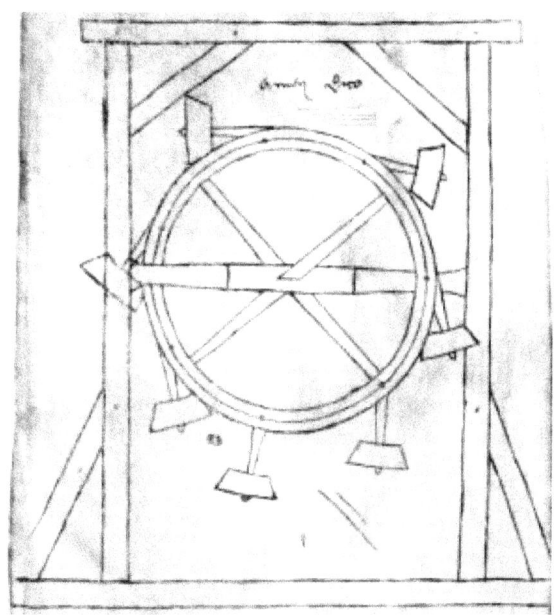

Perpetuum Mobile of Villard de Honnecourt (about 1230).

Gravity also acts at a distance, without an apparent energy source, but to get energy out of a gravitational field (for instance, by dropping a heavy object, producing kinetic energy as it falls) one has to put energy in (for instance, by lifting the object up), and some energy is always dissipated in the process. A typical application of gravity in a perpetual motion machine is Bhaskara's wheel in the 12th cen-

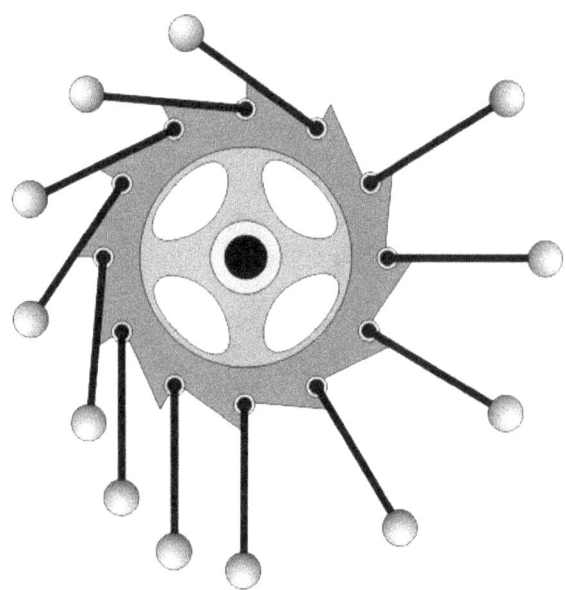

The "Overbalanced Wheel".

tury, whose key idea is itself a recurring theme, often called the overbalanced wheel: moving weights are attached to a wheel in such a way that they fall to a position further from the wheel's center for one half of the wheel's rotation, and closer to the center for the other half. Since weights further from the center apply a greater torque, it was thought that the wheel would rotate forever. However, since the side with weights further from the center has fewer weights than the other side, at that moment, the torque is balanced and perpetual movement is not achieved.[20] The moving weights may be hammers on pivoted arms, or rolling balls, or mercury in tubes; the principle is the same.

Another theoretical machine involves a frictionless environment for motion. This involves the use of diamagnetic or electromagnetic levitation to float an object. This is done in a vacuum to eliminate air friction and friction from an axle. The levitated object is then free to rotate around its center of gravity without interference. However, this machine has no practical purpose because the rotated object cannot do any work as work requires the levitated object to cause motion in other objects, bringing friction into the problem. Furthermore, a *perfect* vacuum is an unattainable goal since both the container and the object itself would slowly vaporize, thereby degrading the vacuum.

To extract work from heat, thus producing a perpetual motion machine of the second kind, the most common approach (dating back at least to Maxwell's demon) is *unidirectionality*. Only molecules moving fast enough and in the right direction are allowed through the demon's trap door. In a Brownian ratchet, forces tending to turn the ratchet one way are able to do so while forces in the other direction are not. A diode in a heat bath allows through currents in one

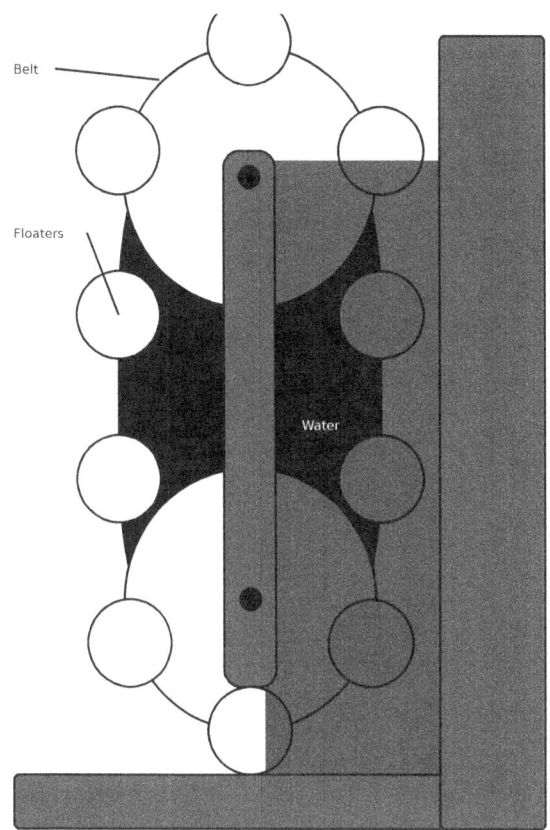

Perpetual motion wheels from a drawing of Leonardo da Vinci.

The "Float Belt". The yellow blocks indicate floaters. It was thought that the floaters would rise through the liquid and turn the belt. However, pushing the floaters into the water at the bottom takes as much energy as the floating generates, and some energy is dissipated.

direction and not the other. These schemes typically fail in two ways: either maintaining the unidirectionality costs energy (requiring Maxwell's demon to perform more thermodynamic work to gauge the speed of the molecules than the amount of energy gained by the difference of temperature caused) or the unidirectionality is an illusion and occasional big violations make up for the frequent small non-violations (the Brownian ratchet will be subject to internal Brownian forces and therefore will sometimes turn the wrong way).

Buoyancy is another frequently misunderstood phenomenon. Some proposed perpetual-motion machines miss the fact that to push a volume of air down in a fluid takes the same work as to raise a corresponding volume of fluid up against gravity. These types of machines may involve two chambers with pistons, and a mechanism to squeeze the air out of the top chamber into the bottom one, which then becomes buoyant and floats to the top. The squeezing mechanism in these designs would not be able to do enough work to move the air down, or would leave no excess work available to be extracted.

1.3 Patents

Proposals for such inoperable machines have become so common that the United States Patent and Trademark Office (USPTO) has made an official policy of refusing to grant patents for perpetual motion machines without a working model. The USPTO Manual of Patent Examining Practice states:

> With the exception of cases involving perpetual motion, a model is not ordinarily required by the Office to demonstrate the operability of a device. If operability of a device is questioned, the applicant must establish it to the satisfaction of the examiner, but he or she may choose his or her own way of so doing.[21]

And, further, that:

> A rejection [of a patent application] on the

ground of lack of utility includes the more specific grounds of inoperativeness, involving perpetual motion. A rejection under 35 U.S.C. 101 for lack of utility should not be based on grounds that the invention is frivolous, fraudulent or against public policy.[22]

The filing of a patent application is a clerical task, and the USPTO will not refuse filings for perpetual motion machines; the application will be filed and then most probably rejected by the patent examiner, after he has done a formal examination.[23] Even if a patent is granted, it does not mean that the invention actually works, it just means that the examiner believes that it works, or was unable to figure out why it would not work.[23]

The USPTO maintains a collection of Perpetual Motion Gimmicks.

The United Kingdom Patent Office has a specific practice on perpetual motion; Section 4.05 of the UKPO Manual of Patent Practice states:

> Processes or articles alleged to operate in a manner which is clearly contrary to well-established physical laws, such as perpetual motion machines, are regarded as not having industrial application.[24]

Examples of decisions by the UK Patent Office to refuse patent applications for perpetual motion machines include:[25]

> Decision BL O/044/06, John Frederick Willmott's application no. 0502841[26]
>
> Decision BL O/150/06, Ezra Shimshi's application no. 0417271[27]

The European Patent Classification (ECLA) has classes including patent applications on perpetual motion systems: ECLA classes "F03B17/04: Alleged perpetua mobilia ..." and "F03B17/00B: [... machines or engines] (with closed loop circulation or similar : ... Installations wherein the liquid circulates in a closed loop; Alleged perpetua mobilia of this or similar kind ...".[28]

1.4 Apparent perpetual motion machines

As "perpetual motion" can only exist in isolated systems, and true isolated systems do not exist, there are not any real "perpetual motion" devices. However, there are concepts and technical drafts that propose "perpetual motion",

but on closer analysis it is revealed that they actually "consume" some sort of natural resource or latent energy, such as the phase changes of water or other fluids or small natural temperature gradients, or simply cannot sustain indefinite operation. In general, extracting work from these devices is impossible.

1.4.1 Resource consuming

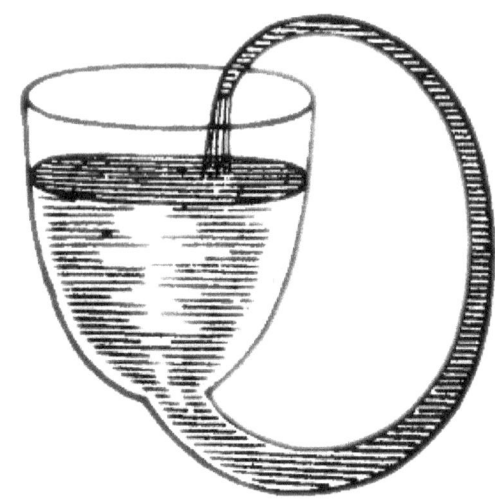

The "Capillary Bowl"

Some examples of such devices include:

- The drinking bird toy functions using small ambient temperature gradients and evaporation. It runs until all water is evaporated.

- A capillary action-based water pump functions using small ambient temperature gradients and vapour pressure differences. With the "Capillary Bowl", it was thought that the capillary action would keep the water flowing in the tube, but since the cohesion force that draws the liquid up the tube in the first place holds the droplet from releasing into the bowl, the flow is not perpetual.

- A Crookes radiometer consists of a partial vacuum glass container with a lightweight propeller moved by (light-induced) temperature gradients.

- Any device picking up minimal amounts of energy from the natural electromagnetic radiation around it, such as a solar powered motor.

- Any device powered by changes in air pressure, such as some clocks (Cox's timepiece, Beverly Clock). The motion leeches energy from moving air which in turn gained its energy from being acted on.

- The Atmos clock uses changes in the vapor pressure of ethyl chloride with temperature to wind the clock spring.

- A device powered by radioactive decay from an isotope with a relatively long half-life; such a device could plausibly operate for hundreds or thousands of years.

- The Oxford Electric Bell and Karpen Pile driven by dry pile batteries.

1.4.2 Low friction

- In flywheel energy storage, "modern flywheels can have a zero-load rundown time measurable in years".[29]

- Once spun up, objects in the vacuum of space—stars, black holes, planets, moons, spin-stabilized satellites, etc.—continue spinning almost indefinitely with no further energy input. Tides on Earth are dissipating the gravitational energy of the Moon/Earth system at an average rate of about 3.75 terawatts.[30][31]

- In certain quantum-mechanical systems (such as superfluidity and superconductivity), dissipation-free "motion" is possible.

1.4.3 Thought experiments

In some cases a thought (or *gedanken*) experiment appears to suggest that perpetual motion may be possible through accepted and understood physical processes. However, in all cases, a flaw has been found when all of the relevant physics is considered. Examples include:

- Maxwell's Demon: This was originally proposed to show that the Second Law of Thermodynamics applied in the statistical sense only, by postulating a "demon" that could select energetic molecules and extract their energy. Subsequent analysis (and experiment) have shown there is no way to physically implement such a system that does not result in an overall increase in entropy.

- Brownian Ratchet: In this thought experiment, one imagines a paddle wheel connected to a ratchet. Brownian motion would cause surrounding gas molecules to strike the paddles, but the ratchet would only allow it to turn in one direction. A more thorough analysis showed that when a physical ratchet was considered at this molecular scale, Brownian motion would also affect the ratchet and cause it to randomly fail resulting in no net gain. Thus, the device would not violate the Laws of thermodynamics.

- Vacuum energy and Zero-point energy: In order to explain effects such as virtual particles and the Casimir effect, many formulations of quantum physics include a background energy which pervades empty space, known as the vacuum energy. More generally, because systems are only allowed to occupy certain discrete energy levels, the lowest energy state of a system is usually greater than zero, and this is known as the "zero-point energy". Inventors have proposed various methods for extracting useful work from the zero-point energy of a system or vacuum, but none have been found to be viable.[32][33] Even though the zero-point energy is theoretically infinite, there is as yet no evidence to suggest that infinite amounts of zero-point energy are available for use, that zero-point energy can be withdrawn for free, or that zero-point energy can be used in violation of conservation of energy.[34]

1.5 See also

- Incredible utility
- Pathological science

1.6 References

[1] Angrist, Stanley (January 1968). "Perpetual Motion Machines". *Scientific American* **218** (1): 115–122.

[2] Oxlade, Chris (2006). *Friction And Resistance*. Heinemann-Raintree Library. p. 27. ISBN 1403481717.

[3] Derry, Gregory N. *What Science Is and How It Works*. Princeton University Press. p. 167. ISBN 1400823110.

[4] Roy, Bimalendu Narayan (2002). *Fundamentals of Classical and Statistical Thermodynamics*. John Wiley & Sons. p. 58. ISBN 0470843136.

[5] "Definition of perpetual motion". Oxforddictionaries.com. 2012-11-22. Retrieved 2012-11-27.

[6] Taylor, J. H.; Weisberg, J. M. (1989). "Further experimental tests of relativistic gravity using the binary pulsar PSR 1913 + 16". *Astrophysical Journal* **345**: 434–450. Bibcode:1989ApJ...345..434T. doi:10.1086/167917.

[7] Weisberg, J. M.; Nice, D. J.; Taylor, J. H. (2010). "Timing Measurements of the Relativistic Binary Pulsar PSR B1913+16". *Astrophysical Journal* **722**: 1030–1034. arXiv:1011.0718v1. Bibcode:2010ApJ...722.1030W. doi:10.1088/0004-637X/722/2/1030.

[8] Simanek, Donald E. (2012). "Perpetual Futility: A short history of the search for perpetual motion". *The Museum of Unworkable Devices*. Donald Simanek's website, Lock Haven University. Retrieved 3 October 2013.

[9] quote originally from Leonardo's notebooks, South Kensington Museum MS *ii* p. 92 McCurdy, Edward (1906). *Leonardo da Vinci's note-books.* US: Charles Scribner's Sons. p. 64.

[10] Rao, Y. V. C. (2004). *An Introduction to Thermodynamics.* Hyderabad, India: Universities Press (India) Private Ltd. ISBN 81-7371-461-4. Retrieved August 2010.

[11] An alternative definition is given, for example, by Schadewald, who defines a "perpetual motion machine of the third kind" as a machine that violates the third law of thermodynamics. See Schadewald, Robert J. (2008), Worlds of Their Own - A Brief History of Misguided Ideas: Creationism, Flat-Earthism, Energy Scams, and the Velikovsky Affair, Xlibris, ISBN 978-1-4363-0435-1. pp55–56

[12] Wong, Kau-Fui Vincent (2000). *Thermodynamics for Engineers.* CRC Press. p. 154. ISBN 978-0-84-930232-9.

[13] Akshoy, Ranjan Paul; Sanchayan, Mukherjee; Pijush, Roy (2005). *Mechanical Sciences: Engineering Thermodynamics and Fluid Mechanics.* Prentice-Hall India. p. 51. ISBN 978-8-12-032727-6.

[14] Barrow, John D. (1998). *Impossibility: The Limits of Science and the Science of Limits.* Oxford University Press. ISBN 978-0-19-851890-7.

[15] Goldstein, Herbert; Poole, Charles; Safko, John (2002). "Classical Mechanics (3rd edition)". San Francisco: Addison Wesley: 589–598. ISBN 0-201-65702-3.

[16] "The perpetual myth of free energy". *BBC News.* 9 July 2007. Retrieved 16 August 2010. In short, law states that energy cannot be created or destroyed. Denying its validity would undermine not just little bits of science - the whole edifice would be no more. All of the technology on which we built the modern world would lie in ruins.

[17] "CE410: Are constants constant?", talkorigins

[18] Harmor, Greg; Derek Abbott (2005). "The Feynman-Smoluchowski ratchet". *Parrondo's Paradox Research Group.* School of Electrical & Electronic Engineering, Univ. of Adelaide. Retrieved 2010-01-15.

[19] Dircks, Henry (1861). *Perpetuum Mobile: Or, A History of the Search for Self-motive.* p. 354. Retrieved 17 August 2012.

[20] Jenkins, Alejandro (2013). "Self-oscillation". *Physics Reports* **525** (2): 167–222. arXiv:1109.6640. Bibcode:2013PhR...525..167J. doi:10.1016/j.physrep.2012.10.007.

[21] "600 Parts, Form, and Content of Application - 608.03 Models, Exhibits, Specimens". *Manual of Patent Examining Procedure* (8 ed.). August 2001.

[22] "700 Examination of Applications II. UTILITY - 706.03(a) Rejections Under 35 U.S.C. 101". *Manual of Patent Examining Procedure* (8 ed.). August 2001.

[23] Pressman, David (2008). Nolo, ed. *Patent It Yourself* (13, illustrated, revised ed.). Nolo. p. 99. ISBN 1-4133-0854-6.

[24] "Manual of Patent Practice, Section 4" (PDF). United Kingdom Patent Office.

[25] See also, for more examples of refused patent applications at the United Kingdom Patent Office (UK-IPO), *UK-IPO gets tougher on perpetual motion*, IPKat, 12 June 2008. Consulted on June 12, 2008.

[26] "Patents Ex parte decision (O/044/06)" (PDF). Retrieved 2013-03-04.

[27] http://www.patent.gov.uk/patent/p-decisionmaking/p-challenge/p-challenge-decision-results/o15006.pdf

[28] ECLA classes F03B17/04 and F03B17/00B. Consulted on June 12, 2008.

[29] WO application 2008037004, Kwok, James, "An energy storage device and method of use", published 2008-04-03

[30] Munk, W.; Wunsch, C (1998). "Abyssal recipes II: energetics of tidal and wind mixing". *Deep Sea Research Part I Oceanographic Research Papers* **45** (12): 1977. Bibcode:1998DSRI...45.1977M. doi:10.1016/S0967-0637(98)00070-3.

[31] Ray, R. D.; Eanes, R. J.; Chao, B. F. (1996). "Detection of tidal dissipation in the solid Earth by satellite tracking and altimetry". *Nature* **381** (6583): 595. Bibcode:1996Natur.381..595R. doi:10.1038/381595a0.

[32] Martin Gardner, "'Dr' Bearden's Vacuum Energy", *Skeptical Inquirer*, January/February 2007

[33] Amber M. Aiken, Ph.D. "Zero-Point Energy: Can We Get Something From Nothing?" (PDF). U.S. Army National Ground Intelligence Center. Forays into "free energy" inventions and perpetual-motion machines using ZPE are considered by the broader scientific community to be pseudoscience.

[34] "FOLLOW-UP: What is the 'zero-point energy' (or 'vacuum energy') in quantum physics? Is it really possible that we could harness this energy?". *Scientific American*. 18 August 1997.

1.7 External links

- Perpetual motion at DMOZ
- The Museum of Unworkable Devices
- Vlatko Vedral's Lengthy discussion of Maxwell's Demon (PDF)
- "Perpetual Motion - Just Isn't." *Popular Mechanics*, January 1954, pp. 108–111.

- In Our Time: Perpetual Motion, BBC discussion with Ruth Gregory, Frank Close and Steven Bramwell, hosted by Melvyn Bragg, first broadcast 24 September 2015.

Chapter 2

History of perpetual motion machines

An engraving of Robert Fludd's 1618 "water screw" perpetual motion machine.

The history of perpetual motion machines dates back to the Middle Ages. For millennia, it was not clear whether perpetual motion devices were possible or not, but the development of modern theories of thermodynamics has shown that they are impossible. Despite this, many attempts have been made to construct such machines, continuing into modern times. Modern designers and proponents often use other terms, such as "overunity", to describe their inventions.

2.1 History

2.1.1 Pre-19th century

The "magic wheel", a wheel spinning on its axle powered by lodestones, appeared in 8th-century Bavaria. The wheel was supposed to rotate perpetually; in fact, it did rotate for a long time, but friction inevitably eventually stopped it.[1] Early designs of perpetual motion machines were done by Indian mathematician–astronomer Bhaskara II, who described a wheel (Bhāskara's wheel) that he claimed would run forever.[2]

A drawing of a perpetual motion machine appeared in the sketchbook of Villard de Honnecourt, a 13th-century French master mason and architect. The sketchbook was concerned with mechanics and architecture. Following the example of Villard, Peter of Maricourt designed a magnetic globe which, if it were mounted without friction parallel to the celestial axis, would rotate once a day. It was intended to serve as an automatic armillary sphere.[2]

Leonardo da Vinci made a number of drawings of devices he hoped would make free energy. Leonardo da Vinci was generally against such devices, but drew and examined numerous overbalanced wheels.[3][4]

Mark Anthony Zimara, a 16th-century Italian scholar, proposed a self-blowing windmill.[5]

Various scholars in this period investigated the topic. In 1607 Cornelius Drebbel in "Wonder-vondt van de eeuwighe beweging" dedicated a Perpetuum motion machine to James I of England.[6] It was described by Heinrich Hiesserle von Chodaw in 1621.[7] Robert Boyle devised the "perpetual vase" ("perpetual goblet" or "hydrostatic paradox") which was discussed by Denis Papin in the *Philosophical Transactions* for 1685.[8] Johann Bernoulli proposed a fluid energy machine. In 1686, Georg Andreas Böckler, designed a "self operating" self-powered water mill and several perpetual motion machines using balls using variants of Archimedes' screws. In 1712, Johann

Bessler (Orffyreus), investigated 300 different perpetual motion models and claimed he had the secret of perpetual motion.

In the 1760s, James Cox and John Joseph Merlin developed Cox's timepiece.[9] Cox claimed that the timepiece was a true perpetual motion machine, but as the device is powered by changes in atmospheric pressure via a mercury barometer, this is not the case.

In 1775, the Royal Academy of Sciences in Paris made the statement that the Academy "will no longer accept or deal with proposals concerning perpetual motion."

2.1.2 Industrial Revolution

19th century

In 1812, Charles Redheffer, in Philadelphia, claimed to have developed a "generator" that could power other machines. Upon investigation, it was deduced that the power was being supplied from another connected machine. Robert Fulton exposed Redheffer's schemes during an exposition of the device in New York City (1813). Removing some concealing wooden strips, Fulton found a catgut belt drive went through a wall to an attic. In the attic, a man was turning a crank to power the device.[10][11][12]

In 1827, Sir William Congreve, 2nd Baronet devised a machine running on capillary action that would disobey the law of liquids never rising above their own level, so to produce a continuous ascent and overflow. The device had an inclined plane over pulleys. At the top and bottom, there travelled an endless band of sponge, a bed and, over this, again an endless band of heavy weights jointed together. The whole stood over the surface of still water. Congreve believed his system would operate continuously.

In 1868, an Austrian, Alois Drasch, received a US patent for a machine that possessed a "thrust key-type gearing" of a rotary engine. The vehicle driver could tilt a trough depending upon need. A heavy ball rolled in a cylindrical trough downward, and, with continuous adjustment of the device's levers and power output, Drasch believed that it would be possible to power a vehicle.[13]

In 1870, E.P. Willis of New Haven, Connecticut made money from a "proprietary" perpetual motion machine. A story of the overly complicated device with a hidden source of energy appears in the *Scientific American* article "The Greatest Discovery Ever Yet Made". Investigation into the device eventually found a source of power that drove it.[14]

John Ernst Worrell Keely claimed the invention of an induction resonance motion motor. He explained that he used "etheric technology". In 1872, Keely announced that he had discovered a principle for power production based on the vibrations of tuning forks. Scientists investigated his machine which appeared to run on water, though Keely endeavoured to avoid this. Shortly after 1872, venture capitalists accused Keely of fraud (they lost nearly five million dollars). Keely's machine, it was discovered after his death, was based on hidden air pressure tubes.[15]

1900 to 1950

In 1900, Nikola Tesla claimed to have discovered an abstract principle on which to base a perpetual motion machine of the second kind. No prototype was produced. He wrote:

In the 1910s and 1920s, Harry Perrigo of Kansas City, Missouri, a graduate of MIT, claimed development of a free energy device.[17] Perrigo claimed the energy source was "from thin air" or from aether waves. Perrigo demonstrated the device before the Congress of the United States on December 15, 1917. Perrigo had a pending application[18] for the "Improvement in Method and Apparatus for Accumulating and Transforming Ether Electric Energy". Investigators report that his device contained a hidden motor battery.[19]

Cover of the October 1920 issue of Popular Science *magazine*

Popular Science, in the October 1920 issue, published an article on the lure of perpetual motion.[20]

2.1.3   Modern era

1951 to 1980

During the middle of the 20th century, Viktor Schauberger claimed to have discovered some special vortex energy in water. Since his death in 1958, people are still studying his works.[21]

In 1966, Josef Papp (sometimes referred to as Joseph Papp or Joseph Papf) supposedly developed an alternative car engine that used inert gases. He gained a few investors but when the engine was publicly demonstrated, an explosion killed one of the observers and injured two others. Papp blamed the accident on interference by physicist Richard Feynman, who later shared his observations in an article in *Laser*, the journal of the Southern Californian Skeptics.[22] Papp continued to accept money but never demonstrated another engine.

On December 20 of 1977, Emil T. Hartman received U.S. Patent 4,215,330 titled "Permanent magnet propulsion system". This device is related to the Simple Magnetic Overunity Toy (SMOT).

ter molecules into high-octane gasoline compounds (named Mota fuel) that would reduce the price of gasoline to 8 cents per gallon. This process involved a green powder (this claim may be related to the similar ones of John Andrews (1917)). He was brought to court for fraud in 1954 and acquitted, but in 1973 was convicted. Justice William Bauer and Justice Philip Romiti both observed a demonstration in the 1954 case.[27]

In 1958, Otis T. Carr from Oklahoma formed a company to manufacture UFO-styled spaceships and hovercraft. Carr sold stock for this commercial endeavour. He also promoted free energy machines. He claimed inspiration from Nikola Tesla, among others.[28]

In 1962, physicist Richard Feynman discussed a Brownian ratchet that would supposedly extract meaningful work from Brownian motion, though he went on to demonstrate how such a device would fail to work in practice.[29]

In the 1970s, David Hamel produced the Hamel generator, an "antigravity" device, supposedly after an alien abduction. The device was tested on *MythBusters* where it failed to demonstrate any lift-generating capability.[30][31]

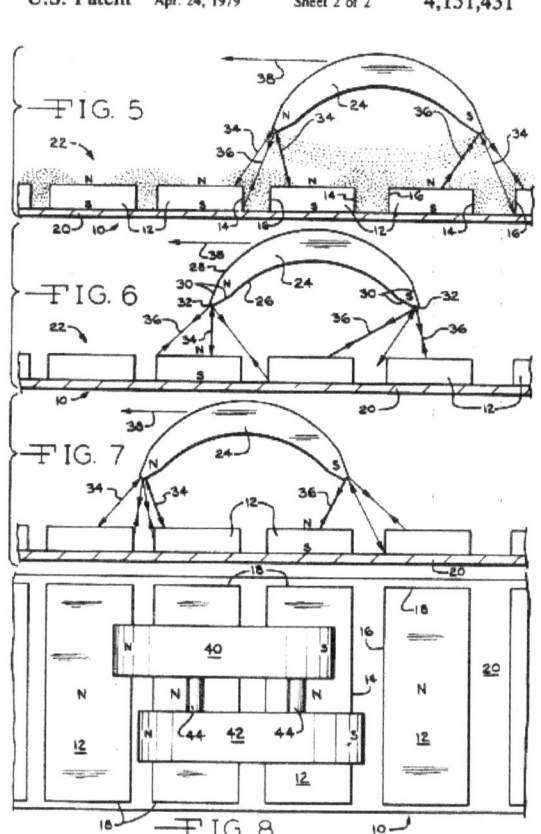

Thesta-Distatica[23] *electrical circuit as explained in Potter's "Methernitha Back-Engineered"*[24] *article.*

Paul Bauman, a German engineer, developed a machine referred to as the "Testatika"[25] and known as the "Swiss M-L converter"[26] or "Thesta-Distatica".[23]

Guido Franch reportedly had a process of transmuting wa-

Howard R. Johnson's US Patent 4151431

Howard Robert Johnson developed a permanent magnet motor and, on April 24, 1979, received U.S. Patent 4,151,431.[The United States Patent office main classification of his 4151431 patent is as a "electrical generator or motor structure, dynamoelectric, linear" (310/12).] Johnson said that his device generates motion, either rotary or linear, from nothing but permanent magnets in rotor as well as stator, acting against each other.[32] He estimated that permanent magnets made of proper hard materials should lose less than two percent of their magnetization in powering a device for 18 years.[33]

1981 to 1999

Dr. Yuri S. Potapov of Moldova claims development of an over-unity electrothermal water-based generator (referred to as "Yusmar 1"). He founded the YUSMAR company to promote his device. The device has failed to produce over unity under tests.[34][35]

CETI claimed development of a device that outputs small yet anomalous amounts of heat, perhaps due to cold fusion. Skeptics state that inaccurate measurements of friction effects from the cooling flow through the pellets may be responsible for the results.[36]

2000s

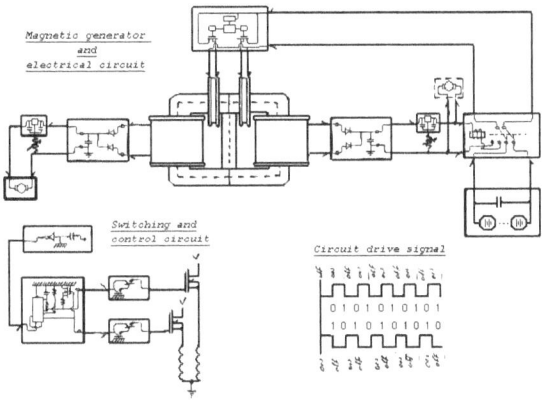

Motionless electromagnetic generator circuit as explained in US Patent 6362718

The motionless electromagnetic generator (MEG) was built by Tom Bearden. Allegedly, the device can eventually sustain its operation in addition to powering a load without application of external electrical power. Bearden claimed that it didn't violate the first law of thermodynamics because it extracted vacuum energy from the immediate environment.[37] Critics dismiss this theory and instead identify it as a perpetual motion machine with an unscientific rationalization.[37][38][39][40][41] Science writer Martin Gardner said that Bearden's physics theories, compiled in the self-published book *Energy from the Vacuum*, are considered "howlers" by physicists, and that his doctorate title was obtained from a diploma mill.[37] Bearden then founded and directed the Alpha Foundation's Institute for Advanced Study (AIAS) to further propagate his theories. This group has published papers in established physics journals and in books published by leading publishing houses, but one analysis lamented these publications because the texts were "full of misconceptions and misunderstandings concerning the theory of the electromagnetic field."[42] When Bearden was awarded U.S. Patent 6,362,718 in 2002, skeptic Robert Park complained so loudly that the American Physical Society issued a statement against the granting.[40] The United States Patent and Trademark Office said that it would reexamine the patent and change the way it recruits examiners, and re-certify examiners on a regular basis, to prevent similar patents from being granted again.[43]

In 2002, the GWE (Genesis World Energy) group claimed to have 400 people developing a device that supposedly separated water into H_2 and O_2 using less energy than conventionally thought possible. No independent confirmation was ever made of their claims, and in 2006, company founder Patrick Kelly was sentenced to five years in prison for stealing funds from investors.[44]

In 2006, Steorn Ltd. claimed to have built an over-unity device based on rotating magnets, and took out an advertisement soliciting scientists to test their claims. The selection process for twelve began in September 2006 and concluded in December 2006.[45] The selected jury started investigating Steorn's claims. A public demonstration scheduled for July 4, 2007 was canceled due to "technical difficulties".[46] In June 2009, the selected jury said the technology does not work.[47]

2.2 See also

- History of science

2.3 References

[1] Mark E. Eberhart(2007):*Feeding the fire: the lost history and uncertain future of mankind's energy*,p.14

[2] Lynn Townsend White, Jr. (April 1960). "Tibet, India, and Malaya as Sources of Western Medieval Technology", *The American Historical Review* **65** (3), p. 522-526.

[3] Time-Life Books (1991). *Inventive Genius*. 143 pages. Page 125. ISBN 0-8094-7699-1

[4] Philip J. Mirowski, (1991). *More Heat Than Light: Economics As Social Physics: Physics As Nature's Economics* 462 pages. Page 15.

[5] Ord-Hume, Arthur W. J. G. (1977). *Perpetual Motion: The History of an Obsession*. New York: St. Martin's Press. pp. 41–44. ISBN 0-312-60131-X. Retrieved March 12, 2011.

[6] "Wonder-vondt van de eeuwighe bewegingh" (PDF). Retrieved 2013-03-04.

[7] http://www.drebbel.net/1621%20PPM.pdf

[8] MIT, "Inventor of the Week Archive: Pascal : Mechanical Calculator", May 2003. "Pascal worked on many versions of the devices, leading to his attempt to create a perpetual motion machine. He has been credited with introducing the roulette machine, which was a by-product of these experiments."

[9] Ord-Hume, Arthur W. J. G. Perpetual Motion: The History of an Obsession. St. Martin's Press. ISBN 0-312-60131-X.

[10] Redheffer's Perpetual Motion Machine

[11] Information originally at www.skepticfiles.org/skep2/pmotion2.htm

[12] No. 438: Redheffer's PMM-I by John H. Lienhard

[13] pmm_physics German page0300

[14] Image of Scientific American

[15] "Keely's Secret Disclosed.; Scientists Examine His Laboratory and Discover Hidden Tubes in Proof of His Deception." (PDF), *New York Times*, 20 January 1899

[16] N. Tesla, "The problem of increasing human energy: With special references to the harnessing of the sun's energy", *Century Magazine* (1900). Full text available here

[17] Harry E. Perrigo, a vertical file at the Kansas City Public Library in Kansas City, Missouri, described as follows: "Photos, illustrations, and information on Harry Perrigo, a local inventor of a "free energy" device in the 1910s-1920s turning out to be a hoax. Energy source of "invention" supposedly "from thin air" or from "ether waves" but in actually from a hidden battery."

[18] filed December 31, 1925; Serial Number 78,719

[19] Ciations originally at www.kclibrary.org resources Subject area ID 77176

[20] Phillip Rowland. "The undying lure of perpetual motion". *Popular Science*, October 1920.

[21] Who was Viktor Schauberger? frank.germano.com. (archived version).

[22] R. Feynman on Papp perpetual motion engine; Originally published in LASER, Journal of the Southern Californian Skeptics

[23] New scientist, Volume 170, Issues 2286-2291. Page 48.

[24] "Back-Engineered Testatika" by Paul E. Potter

[25] Jeane Manning, The coming energy revolution. Reviews Avery Pub. Group, April 1, 1996. Page 141.

[26] Matthey, PH (1985), "The Swiss ML Converter - A Masterpiece of Craftsmanship and Electronic Engineering"

[27] The Straight Dope: Is there a pill that can turn water into gasoline?

[28] Otis Carr Flying Machine – KeelyNet 12/23/01

[29] Feynman, Richard P. (1963). *The Feynman Lectures on Physics, Vol. 1*. Massachusetts, USA: Addison-Wesley. Chapter 46. ISBN 0-201-02116-1.

[30] http://kwc.org/mythbusters/2006/12/episode_68_christmas_tree_ligh.html

[31] MythBusters Episode 68: Christmas Tree Lights, Antigravity Device, Vodka Myths IV

[32] Jorma Hyypia (Spring 1980). "Amazing Magnet-Powered Motor". *Science & Mechanics*. (Cover illustration is here.)

[33] "Industrial Engineer Gets Patent for a Device Powered by Magnets". *Sarasota Herald-Tribune*. April 29, 1979.

[34] ETI – Experiments

[35] Commercial Sources

[36] CETI : Patterson Cell – taking a scientific look

[37] Martin Gardner, "'Dr' Bearden's Vacuum Energy", *Skeptical Inquirer*, January/February 2007

[38] "Free Energy: Perpetual Motion Scams are at an All-Time High", *What's New*, APS, 5 April 2002

[39] "Vacuum Energy: How Do You Patent a Perpetual-Motion Machine?", *What's New*, APS, 3 May 2002

[40] "Free Energy: APS Board Speaks Out on Perpetual Motion", *What's New*, APS, 28 June 2002 "The Executive Board of the American Physical Society is concerned that in this period of unprecedented scientific advance, misguided or fraudulent claims of perpetual motion machines and other sources of unlimited free energy are proliferating. Such devices would directly violate the most fundamental laws of Nature, laws that have guided the scientific advances that are transforming our world."

[41] Energy: No Such Thing as a Free Lunch", Eric Prebys, Guest Lecture at Columbia University, November 4, 2008. Also, shorter version for Fermilab's "Ask a scientist" program, December 6, 2009

[42] Trovon De Carvalho, A. L.; Rodrigues, W. A. (July 15, 2003). "The *non-sequitur* mathematics and physics of the New Electrodynamics proposed by the AIAS group" (PDF). *Random Operators and Stochastic Equations* (Walter de Gruyter GmbH & Co.) **9** (2): 161–206. Retrieved 24 March 2013. We show that the AIAS group collection of papers on a 'new electrodynamics' recently published in the Journal of New Energy, as well as other papers signed by that group (and also other authors) appearing in other established physical journals and in many books published by leading international publishers (see references) are full of misconceptions and misunderstandings concerning the theory of the electromagnetic field and contain fatal mathematical flaws, which invalidates almost all claims done by the authors.

[43] "Free Energy: The Patent Office Decides to Take Another Look", *What's New*, APS, 23 August 2002

[44] State of New Jersey

[45] Originally at http://steorn.net/en/news.aspx?p=2&id=911

[46] Originally at http://steorn.net/en/news.aspx?p=2&id=981

[47] "Irish 'energy for nothing' gizmo fails jury vetting". *The Irish Times*. June 6, 2009.

2.4   Further reading

- Childress, David Hatcher (1994). *The Free-Energy Device Handbook: A Compilation of Patents & Reports.* Stelle, Ill: Adventures Unlimited Press. ISBN 978-0932813244

- Dircks, Henry. (1870). *Perpetuum Mobile: Or, A History of the Search For Self-Motive Power, From the 13th to the 19th Century* With an introductory essay. Second Series. London. W. Clowes and Sons

- Verance, Percy. (1916). *Perpetual Motion: Comprising a History of the Efforts to Attain Self-Motive Mechanism with a Classified, Illustrated Collection and Explanation of the Devices Whereby it Has Been Sought and Why They Failed, and Comprising Also a Revision and Re-Arrangement of the Information Afforded by "Search for Self -Motive Power During The 17th, 18th and 19th Centuries," London, 1861, and "A History of the Search for Self-Motive Power from the 13th to The 19th Century," London, 1870, by Henry Dircks, C. E., LL. D., Etc..* 20th Century Enlightenment Specialty Co.

- Ord-Hume, Arthur W. J. G. (1977). Perpetual Motion: The History of an Obsession. St. Martin's Press. ISBN 0-312-60131-X.

- Angrist, Stanley W., *"Perpetual Motion Machines"*. Scientific American. January 1968.

- Hans-Peter, *"Perpetual Motion Chronology"*. HP's Perpetuum Mobile.

- MacMillan, David M., et al., *"The Rolling Ball Web, An Online Compendium of Rolling Ball Sculptures, Clocks, Etc"*.

- Lienhard, John H., *"Perpetual motion"*. The Engines of Our Ingenuity, 1997.

- *"Patents for Unworkable Devices"*. The Museum of Unworkable Devices.

- *"Perpetual Motion Pioneers (The Movers and Shakers)"*. The Museum of Unworkable Devices.

- Boes, Alex, *"Museum of Hoaxes"*.

- Kilty, Kevin T., *"Perpetual Motion"*. 1999.

- The Basement Mechanic's Guide to Testing Perpetual Motion Machines

- 1911 Encyclopedia, *"perpetual motion"*. LoveToKnow, Corp.

2.5   External links

- Allan, Sterling D., *"Free Energy Inventors"*. December 11, 2003.

- Gousseva, Maria, *"Alleged Creation of Perpetual Energy Source Splits Scientific Community"*. Pravda.ru.

- Bearden, Tom, *"Perpetual motion vs. "working machines creating energy from nothing""*. 2003, Revised 2004.

- Perpetuum mobile page by Veljko Milković.

Chapter 3

Johann Bessler

Johann Bessler (Orffyreus)

Johann Ernst Elias Bessler (*ca.* 1680 – November 30, 1745), known as **Orffyreus** or **Orffyré**, was a German entrepreneur who claimed to have built several perpetual motion machines. Those claims generated considerable interest and controversy among some of the leading natural philosophers of the day, including Gottfried Wilhelm Leibniz, Johann Bernoulli, John Theophilus Desaguliers, and Willem 's Gravesande. The modern scientific consensus is that Bessler perpetrated a deliberate fraud, although the details of this have not been satisfactorily explained.

3.1 Life and career

Bessler was born to a peasant family in Upper Lusatia, Germany, *circa* 1680.[1] He completed secondary schooling in Zittau and then travelled widely. An alchemist instructed him on the fabrication of elixirs and he found work as a healer. He was also an apprentice watchmaker until his fortunes improved when he married a wealthy woman in Annaberg.[2]

Bessler adopted the pseudonym "Orffyreus" by writing the letters of the alphabet in a circle and selecting the letters diametrically opposite to those of his surname (what would modernly be called a ROT13 cipher), thus obtaining *Orffyre*, which he then Latinized into *Orffyreus*.[3] That was the name by which he was generally known thereafter.

3.1.1 Orffyreus's wheels

In 1712, Bessler appeared in the town of Gera in the province of Reuss and exhibited a "self-moving wheel," which was about 6½ ft (2 m) in diameter and 4 inches (10 cm) thick. Once in motion it was capable of lifting several pounds.[3] Bessler then moved to Draschwitz, a village near Leipzig, where in 1713 he constructed an even larger wheel, a little over nine feet (2¾ m) in diameter and six inches (15 cm) in width. That wheel could turn at fifty revolutions a minute and raise a weight of forty pounds (18 kg).[3]

The eminent mathematician Gottfried Wilhelm Leibniz visited Draschwitz in 1714 and witnessed a demonstration of Bessler's wheel. In a letter to Robert Erskine, physician and advisor to Russian Tsar Peter the Great, Leibniz later wrote that Bessler was "one of my friends" and that he believed Bessler's wheel to be a valuable invention. Bessler also received support from other members of Leibniz's intellectual circle, including mathematician Johann Bernoulli, philosopher Christian Wolff, and architect Joseph Emanuel Fischer von Erlach.[2]

Bessler then constructed a still larger wheel in Merseburg, before moving to the independent state of Hesse-Kassel,

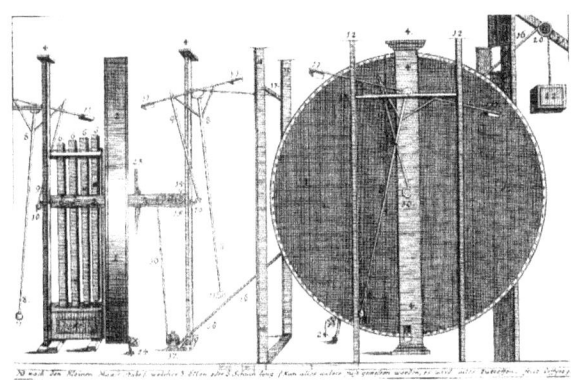

Orffyreus Wheel diagram. (Merseburg, Germany)

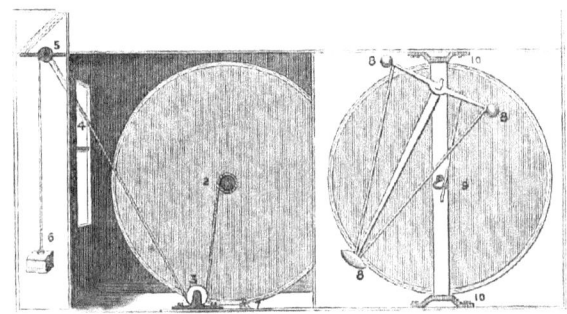

Orffyreus Wheel diagram. (Kassel, Germany)

where Prince Karl, the reigning Landgrave and an enthusiastic patron of mechanical inventors, appointed him as a commercial councillor (*Kommerzialrat*) for the town of Kassel and provided him with rooms in Weissenstein Castle.[4] It was there that in 1717 he constructed his largest wheel so far, 12 feet (3.7 m) in diameter and 14 inches (36 cm) thick.[3]

The inventor demonstrated the operation of his wheel before various audiences, always taking care that the mechanism within the wheel should remain hidden from view, purportedly to prevent others from stealing his invention. The wheel was examined externally by several scientists, including Willem 's Gravesande, professor of mathematics and astronomy at Leiden University, who reported that he could not detect any fraud regarding its operation. On November 12, 1717, the wheel was locked in a room in the castle with the doors and windows sealed to prevent any interference. This was witnessed by the Landgrave and various officials. Two weeks later, the seals were broken and the room was opened, whereupon the wheel was found to be revolving. The door was resealed until January 4, 1718. The wheel was then found to be turning at twenty-six revolutions per minute.[3]

Bessler demanded £20,000 (equivalent to 100,000 Reichsthalers) in exchange for revealing the secret of his machines. Peter the Great was interested in purchasing

the invention and sought advice on the matter from 's Gravesande and others. 's Gravesande examined the axle of the wheel, concluding that he could see no way in which power could be transmitted to it from the outside. Bessler, however, then smashed the wheel, accusing 's Gravesande of trying to discover the secret of the wheel without paying for it, and declaring that the curiosity of the professor had provoked him.[3]

3.1.2   Later life

Bessler and his machine then vanished into obscurity. It is known that he was rebuilding his machine in 1727 and that 's Gravesande had agreed to examine it again, but it is not known whether it was ever tested. Bessler was apparently arrested in 1733, but by 1738 he was again free and living in an estate in Bad Karlshafen, near Kassel.[2] Bessler died in 1745, aged sixty-five, when he fell to his death from a four-and-a-half-story windmill he was constructing in Fürstenburg.[3]

3.2   Mechanism of Orffyreus's Wheel

Bessler's devices were all hollow wheels, with canvas covering the internal mechanism, that turned on a horizontal axis supported by vertical wooden beams on either side of the wheel.[3][4] Christian Wolff, who viewed the wheel in 1715, wrote that Bessler freely revealed that the device utilized weights of about 4 pounds. Fischer von Erlach, who viewed the wheel in 1721, reported: "At every turn of the wheel can be heard the sound of about eight weights, which fall gently on the side toward which the wheel turns."[5] In a letter to Sir Isaac Newton, 's Gravesande reported that, when pushed, the wheel took two or three revolutions to reach a maximum speed of about 25 revolutions per minute.[6] The wheels at Merseburg and Kassel were attached to three-bobbed pendula, one on either side, which presumably acted as regulators, limiting the maximum speed of revolution.[3]

Bessler never revealed the mechanism that kept his wheel in motion and, according to surviving sources, the Landgrave of Hesse-Kassel was the only person whom he ever allowed to examine the inside of the wheel.[3] In 1719 Bessler published a pamphlet in German and Latin, entitled *The Triumphant Orffyrean Perpetual Motion*, which gives a very vague account of his principles.[7] He indicated that the wheel depended upon weights placed so that they can "never attain equilibrium." This suggests that it was a kind of "over-balanced wheel,"[3] a hypothetical gravity-powered device which is now recognized by physicists as impossible (see perpetual motion).

3.2.1 Allegations of fraud

Most of the people who met him, including supporters such as 's Gravesande, reported that Bessler was eccentric, ill-tempered, and perhaps even insane.[3] From the beginning, Bessler's work generated accusations of fraud from various people, including mining engineer Johann Gottfried Borlach, mathematician Christian Wagner, model-maker Andreas Gärtner, Kassel court tutor Jean-Pierre de Crousaz, and others. Gärtner went on to gain employ as model-master for the Polish royal court and in that capacity he built several devices that reproduced some of the successes of Bessler's public demonstrations, including the locked-room test, but which Gärtner acknowledged as mere trickery.[4]

In November 1727, Bessler's maid, Anne Rosine Mauersbergerin, ran away from Bessler's household and testified under oath that she had turned the machines manually from an adjoining room, alternating in that job with Bessler's wife, his brother Gottfried, and Bessler himself. 's Gravesande refused to accept the maid's testimony, writing that he paid "little attention to what a servant can say about machines". By then, 's Gravesande was embroiled in an academic dispute with members of Isaac Newton's circle about the possibility of gravity-powered perpetual motion, which 's Gravesande persistently defended based partly on his belief that Bessler, though "mad", was not a fraud.[2]

The consensus view of modern scientists is that Bessler was perpetrating a deliberate fraud, though just how his wheel was powered is not entirely clear.[2] According to the writers of *Chambers's Encyclopaedia*, Orffyreus's wheel, "but for its strange effect on 's Gravesande, would have been forgotten long ago".[8] If the maid's confession were true, the testimonies by Prince Karl, 's Gravesande, and others about the conditions in which the wheel was tested and exhibited must be flawed.[3]

3.3 References

[1] A contemporary source says that he was born in "Ullersdorff near Zittau" (today Oldrichov, Czech Republic) : "Bibliothèque Germanique ou Histoire littéraire de l'Allemagne, de la Suisse et des Pays du Nord", XXXIX, 1737, p. 15

[2] Jenkins, A. (2013). "The mechanical career of Councillor Orffyreus, confidence man". *American Journal of Physics* **81** (6): 421–427. arXiv:1301.3097. Bibcode:2013AmJPh..81..421J. doi:10.1119/1.4798617.

[3] Rupert T. Gould, "Orffyreus's Wheel," in *Oddities: A Book of Unexplained Facts*, revised ed., (London: Geoffrey Bles, 1944), pp. 89-116. Reprinted by Kessinger Pub Co., 2003, ISBN 978-0-7661-3620-5.

[4] Simon Schaffer (1995). "The show that never ends: perpetual motion in the early eighteenth century". *British Journal for the History of Science* **28** (2): 157–189. doi:10.1017/s0007087400032957. JSTOR 4027676.

[5] Letter from J. E. Fischer to John Theophilus Desaguliers, 1721. Quoted by R.T. Gould, *Oddities*, p. 96.

[6] Letter by W. 's Gravesande to I. Newton, August 1721, quoted by R.T. Gould, *Oddities*, p. 95.

[7] J. E. E. Bessler ("Orffyreus"), *Das Triumphirende Perpetuum mobile Orffyreanum*, (Kassel, 1719).

[8] "Perpetual Motion", *Chambers's encyclopædia: a dictionary of universal knowledge*, vol. 7, (London: W. and R. Chambers, 1868), pp. 412-415.

3.4 External links

- List of works and e-texts at de.wikisource

Chapter 4

Bhāskara's wheel

Bhāskara's wheel was invented in 1150 by Bhāskara II, an Indian mathematician, in an attempt to create a perpetual motion machine. The wheel consisted of curved or tilted spokes partially filled with mercury.[1] Once in motion, the mercury would flow from one side of the spoke to another, thus forcing the wheel to continue motion.

4.1 References

[1] Lynn Townsend White (April 1960). *Tibet, India, and Malaya as Sources of Western Medieval Technology*. p. 65.

4.2 External links

- Shifting-Mass Overbalanced Wheel

Chapter 5

Brownian ratchet

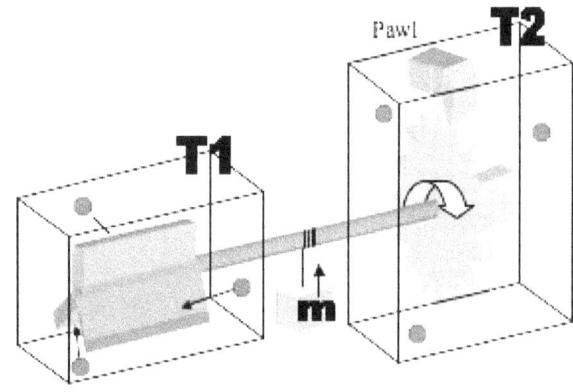

Schematic figure of a Brownian Ratchet

In the philosophy of thermal and statistical physics, the **Brownian ratchet** or **Feynman-Smoluchowski ratchet** is a thought experiment about an apparent perpetual motion machine first analysed in 1912 by Polish physicist Marian Smoluchowski[1] and popularised by American Nobel laureate physicist Richard Feynman in a physics lecture at the California Institute of Technology on May 11, 1962, during his Messenger Lectures series The Character of Physical Law in Cornell University in 1964 and in his text *The Feynman Lectures on Physics*[2] as an illustration of the laws of thermodynamics. The simple machine, consisting of a tiny paddle wheel and a ratchet, appears to be an example of a Maxwell's demon, able to extract useful work from random fluctuations (heat) in a system at thermal equilibrium in violation of the second law of thermodynamics. Detailed analysis by Feynman and others showed why it cannot actually do this.

5.1 The machine

The device consists of a gear known as a ratchet that rotates freely in one direction but is prevented from rotating in the opposite direction by a pawl. The ratchet is connected by an axle to a paddle wheel that is immersed in a fluid of molecules at temperature T_1. The molecules constitute a heat bath in that they undergo random Brownian motion with a mean kinetic energy that is determined by the temperature. The device is imagined as being small enough that the impulse from a single molecular collision can turn the paddle. Although such collisions would tend to turn the rod in either direction with equal probability, the pawl allows the ratchet to rotate in one direction only. The net effect of many such random collisions would seem to be that the ratchet rotates continuously in that direction. The ratchet's motion then can be used to do work on other systems, for example lifting a weight (m) against gravity. The energy necessary to do this work apparently would come from the heat bath, without any heat gradient. Were such a machine to work successfully, its operation would violate the second law of thermodynamics, one form of which states: "It is impossible for any device that operates on a cycle to receive heat from a single reservoir and produce a net amount of work."

5.2 Why it fails

Although at first sight the Brownian ratchet seems to extract useful work from Brownian motion, Feynman demonstrated that if the entire device is at the same temperature, the ratchet will not rotate continuously in one direction but will move randomly back and forth, and therefore will not produce any useful work. The reason is that the pawl, since it is at the same temperature as the paddle, will also undergo Brownian motion, "bouncing" up and down. It therefore will intermittently fail by allowing a ratchet tooth to slip backward under the pawl while it is up. Another issue is that when the pawl rests on the sloping face of the tooth, the spring which returns the pawl exerts a sideways force on the tooth which tends to rotate the ratchet in a backwards direction. Feynman demonstrated that if the temperature T_2 of the ratchet and pawl is the same as the temperature T_1 of the paddle, then the failure rate must equal the rate at which the ratchet ratchets forward, so that no net motion results over long enough periods or in an ensemble averaged sense.[2] A simple but rigorous proof that no net mo-

20

tion occurs no matter what shape the teeth are was given by Magnasco.[3]

If, on the other hand, T_2 is smaller than T_1 , the ratchet will indeed move forward, and produce useful work. In this case, though, the energy is extracted from the temperature gradient between the two thermal reservoirs, and some waste heat is exhausted into the lower temperature reservoir by the pawl. In other words, the device functions as a miniature heat engine, in compliance with the second law of thermodynamics. Conversely, if T_2 is greater than T_1 , the device will rotate in the opposite direction.

The Feynman ratchet model led to the similar concept of Brownian motors, nanomachines which can extract useful work not from thermal noise but from chemical potentials and other microscopic nonequilibrium sources, in compliance with the laws of thermodynamics.[3][4] Diodes are an electrical analog of the ratchet and pawl, and for the same reason cannot produce useful work by rectifying Johnson noise in a circuit at uniform temperature.

Millonas [5] as well as Mahato [6] extended the same notion to correlation ratchets driven by mean-zero (unbiased) nonequilibrium noise with a nonvanishing correlation function of odd order greater than one.

5.3 History

The ratchet and pawl was first discussed as a Second Law-violating device by Gabriel Lippmann in 1900.[7] In 1912, Polish physicist Marian Smoluchowski[1] gave the first correct qualitative explanation of why the device fails; thermal motion of the pawl allows the ratchet's teeth to slip backwards. Feynman did the first quantitative analysis of the device in 1962 using the Maxwell–Boltzmann distribution, showing that if the temperature of the paddle T_1 was greater than the temperature of the ratchet T_2, it would function as a heat engine, but if $T_1 = T_2$ there would be no net motion of the paddle. In 1996, Juan Parrondo and Pep Español used a variation of the above device in which no ratchet is present, only two paddles, to show that the axle connecting the paddles and ratchet conducts heat between reservoirs; they argued that although Feynman's conclusion was correct, his analysis was flawed because of his erroneous use of the quasistatic approximation, resulting in incorrect equations for efficiency.[8] Magnasco and Stolovitzky (1998) extended this analysis to consider the full ratchet device, and showed that the power output of the device is far smaller than the Carnot efficiency claimed by Feynman.[9] A paper in 2000 by Derek Abbott, Bruce R. Davis and Juan Parrondo, reanalyzed the problem and extended it to the case of multiple ratchets, showing a link with Parrondo's paradox.[10]

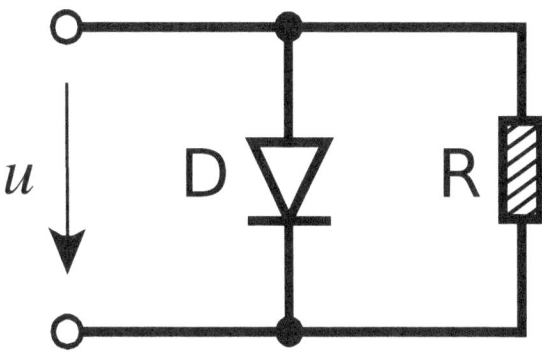

Brillouin paradox: an electrical analogue of the Brownian ratchet.

Léon Brillouin in 1950 discussed an electrical analogy that uses a rectifier (such as a diode) instead of a ratchet.[11] The idea was that the thermal current fluctuations impinging on the diode should be rectified, and therefore spontaneously produce a nonzero constant voltage offset that can be used to perform work. In the detailed analysis it was shown that the thermal fluctuations within the diode generate an electromotive force that cancels the voltage from rectified current fluctuations. For this reason, the diode will only produce a nonzero voltage when the impinging current fluctuations have a different temperature than the diode itself.[12]

5.4 Granular gas

Researchers from the University of Twente, the University of Patras in Greece, and the Foundation for Fundamental Research on Matter have constructed a Feynman-Smoluchowski engine which, when not in thermal equilibrium, converts pseudo-Brownian motion into work by means of a granular gas,[13] which is a conglomeration of solid particles vibrated with such vigour that the system assumes a gas-like state. The constructed engine consisted of four vanes which were allowed to rotate freely in a vibrofluidized granular gas.[14] Because the ratchet's gear and pawl mechanism, as described above, permitted the axle to rotate only in one direction, random collisions with the moving beads caused the vane to rotate. This seems to contradict Feynman's hypothesis. However, this system is not in perfect thermal equilibrium: energy is constantly being supplied to maintain the fluid motion of the beads. Vigorous vibrations on top of a shaking device mimic the nature of a molecular gas. Unlike an ideal gas, though, in which tiny particles move constantly, stopping the shaking would simply cause the beads to drop. In the experiment, this necessary out-of-equilibrium environment was thus maintained. Work was not immediately being done, though; the ratchet effect only commenced beyond a critical shaking strength. For very strong shaking, the vanes of the paddle wheel interacted with the gas, forming a convection roll, sustaining their rotation.[14] The experiment was filmed.

5.5   See also

- Quantum stirring, ratchets, and pumping

- Geometric phase (section Stochastic Pump Effect)

5.6   Notes

[1] M. von Smoluchowski (1912) Experimentell nachweisbare, der Ublichen Thermodynamik widersprechende Molekularphenomene, *Phys. Zeitshur.* **13**, p.1069 cited in Freund, Jan (2000) Stochastic Processes in Physics, Chemistry, and Biology, Springer, p.59

[2] Feynman, Richard P. (1963). *The Feynman Lectures on Physics, Vol. 1.* Massachusetts, USA: Addison-Wesley. Chapter 46. ISBN 0-201-02116-1.

[3] Magnasco, Marcelo O. (1993). "Forced Thermal Ratchets". *Physical Review Letters* **71** (10): 1477–1481. Bibcode:1993PhRvL..71.1477M. doi:10.1103/PhysRevLett.71.1477. PMID 10054418.

[4] Magnasco, Marcelo O. (1994). "Molecular Combustion Motors". *Physical Review Letters* **72** (16): 2656–2659. Bibcode:1994PhRvL..72.2656M. doi:10.1103/PhysRevLett.72.2656. PMID 10055939.

[5] Dante R. Chialvo; Mark Millonas (1995). "Asymmetric unbiased fluctuations are sufficient for the operation of a correlation ratchet". *Physics Letters A* **209** (1-2): 26–30. arXiv:cond-mat/9410057. Bibcode:1995PhLA..209...26C. doi:10.1016/0375-9601(95)00773-0.

[6] M.C. Mahato; A.M. Jayannavar (1995). "ynchronized first-passages in a double-well system driven by an asymmetric periodic field". *Physics Letters A* **209** (1-2): 21–26. Bibcode:1995PhLA..209...21M. doi:10.1016/0375-9601(95)00772-9.

[7] Harmer, Greg; Derek Abbott (2005). "The Feynman-Smoluchowski ratchet". *Parrondo's Paradox Research Group.* School of Electrical & Electronic Engineering, Univ. of Adelaide. Retrieved 2010-01-15.

[8] Parrondo, Juan M. R.; Pep Español (March 8, 1996). "Criticism of Feynman's analysis of the ratchet as an engine". *American Journal of Physics* **64** (9): 1125. Bibcode:1996AmJPh..64.1125P. doi:10.1119/1.18393.

[9] Magnasco, Marcelo O.; Gustavo Stolovitzky (1998). "Feynman's Ratchet and Pawl". *Journal of Statistical Physics* **93** (3): 615. Bibcode:1998JSP....93..615M. doi:10.1023/B:JOSS.0000033245.43421.14.

[10] Abbott, Derek; Bruce R. Davis; Juan M. R. Parrondo (2000). "The problem of detailed balance for the Feynman-Smoluchowski Engine and the multiple pawl paradox" (PDF). *Unsolved Problems of Noise and Fluctuations.* American Institute of Physics. pp. 213–218. Retrieved 2010-01-15.

[11] Brillouin, L. (1950). "Can the Rectifier Become a Thermodynamical Demon?". *Physical Review* **78** (5): 627. Bibcode:1950PhRv...78..627B. doi:10.1103/PhysRev.78.627.2.

[12] Gunn, J. B. (1969). "Spontaneous Reverse Current Due to the Brillouin Emf in a Diode". *Applied Physics Letters* **14** (2): 54. Bibcode:1969ApPhL..14...54G. doi:10.1063/1.1652709.

[13] "Classical thought experiment brought to life in granular gas", *Foundation for Fundamental Research on Matter*, Utrecht, 18 June 2010. Retrieved on 2010-06-24.

[14] Peter Eshuis, Ko van der Weele, Detlef Lohse, and Devaraj van der Meer (June 2010). "Experimental Realization of a Rotational Ratchet in a Granular Gas". *Physical Review Letters* **104** (24): 4. Bibcode:2010PhRvL.104x8001E. doi:10.1103/PhysRevLett.104.248001. PMID 20867337.

5.7   External links

- Why is a Brownian motor not a *perpetuum mobile* of the second kind?

- Coupled Brownian Motors - Can we get work out of unbiased fluctuation?

- Experiment finally proves 100-year-old thought experiment is possible (w/ Video)

- Richard Feynman: Messenger Series lectures videos: hosted by Project Tuva

Articles

- Astumian RD (1997). "Thermodynamics and kinetics of a Brownian motor". *Science* **276** (5314): 917–22. doi:10.1126/science.276.5314.917. PMID 9139648.

- Astumian RD, Hänggi P (2002). "Brownian Motors" (PDF). *Physics Today* **55** (11): 33–9. Bibcode:2002PhT....55k..33A. doi:10.1063/1.1535005.

- Hänggi P, Marchesoni F, Nori F (2005). "Brownian Motors" (PDF). *A. Physik (Leipzig)* **14** (1-3): 51–70. arXiv:cond-mat/0410033. Bibcode:2005AnP...517...51H. doi:10.1002/andp.200410121.

- Lukasz Machura: *Performance of Brownian Motors.* University of Augsburg, 2006 (PDF)

- Peskin CS, Odell GM, Oster GF (July 1993). "Cellular motions and thermal fluctuations: the Brownian ratchet". *Biophys. J.* **65** (1): 316–24. Bibcode:1993BpJ....65..316P. doi:10.1016/S0006-3495(93)81035-X. PMC 1225726. PMID 8369439.

- Hänggi P, Marchesoni F (2008). "Artificial Brownian motors: Controlling transport on the nanoscale: Review" (PDF). arXiv:0807.1283. Bibcode:2009RvMP...81..387H. doi:10.1103/RevModPhys.81.387.

- van Oudensaarden A, Boxer SG (1999). "Brownian Ratchets: Molecular Separations in Lipid Bilayers Supported on Patterned Arrays" (PDF). *Science* **285**: 1046–1048. doi:10.1126/science.285.5430.1046.

Chapter 6

Cox's timepiece

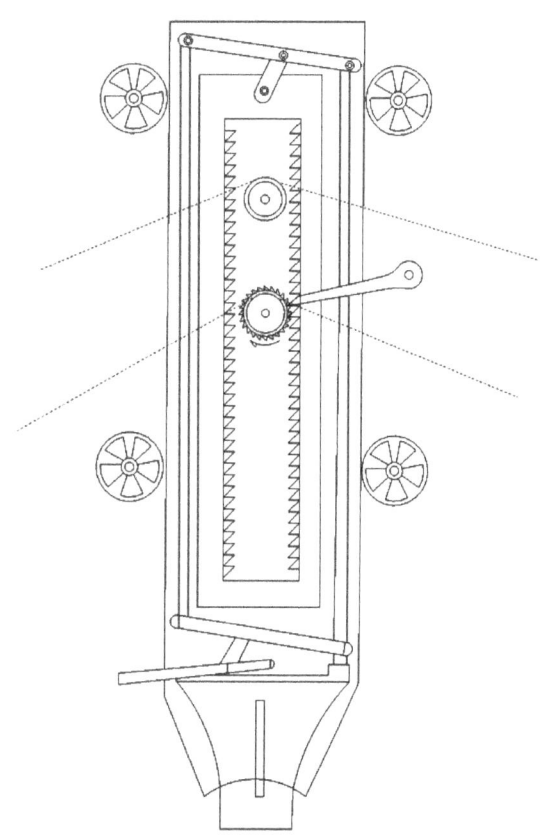

Cox timepiece winding switch

Cox's timepiece is a clock developed in the 1760s by James Cox. It was developed in collaboration with John Joseph Merlin (with whom Cox also worked on developing automata). Cox claimed that his design was a true perpetual motion machine, but as the device is powered from changes in atmospheric pressure via a mercury barometer, this is not the case. The clock still exists today but was deactivated at the time of the clock's relocation to the Victoria and Albert Museum in London.[1]

6.1 Design and history

The clock is similar to other mechanical clocks, except that it does not need winding. The change of pressure in the Earth's atmosphere acts as an external energy source and causes sufficient movement of the winding mechanism. This keeps the mainspring coiled inside the barrel. The clock is designed to enable the timepiece to run indefinitely and overwinding is prevented by a safety mechanism. The prime mover, encased in a finely detailed clock body, is a Fortin mercury barometer. The barometer contained 68 kilograms (150 pounds) of mercury.[2]

Related to this is Cornelis Drebbel's device of 1610 (though it is unknown whether Cox knew of it). It was a machine that told the time, date, and season. The gold machine was mounted in a globe on pillars and was powered by changes in air pressure (a sealed glass tub with liquid varied in volume through atmospheric pressure changes, rewinding constantly).

The Atmos, manufactured by Jaeger LeCoultre is a modern clock which is similar to Cox's clock although the main driving force is generated from temperature differential, instead of pressure differential.

6.2 See also

- History of perpetual motion machines
- Beverly Clock (1864)
- Atmos clock

6.3 References

[1] Ord-Hume, Arthur W. J. G. (1977). *Perpetual Motion: The History of an Obsession*. New York: St. Martin's Press. ISBN 0-312-60131-X., p. 118 (*online copy*, p. 118, at Google Books)

[2] Bruton, Eric (1979). *The History of Clocks and Watches.* New York: Rizzoli International Publications. ISBN 0-8478-0261-2.

6.4 External articles and further reading

6.4.1 Journals

- William Nicholson, "*Concerning those perpetual motions which are produced in machines by the rise and fall of the barometer or thermometrical variations in the dimensions of bodies*". Philosophical Journal.

- William Nicholson, Philosophical Journal, vol I, 1799, p375

6.4.2 Books

- Arthur W. J. G. Ord-Hume: *Perpetual Motion: The History of an Obsession.* Adventures Unlimited Press 2006, ISBN 1-931882-51-7, pp. 110–124 (*online copy*, p. 110, at Google Books)

- Arthur W. J. G. Ord-hume, "*Clockwork Music*", Allen & Unwin, London 1973.

- John Joseph Merlin, "*The Ingenious Mechanick*". The Greater London Council, The Iveagh Bequest, Kenwood, Hampstead Lane, London, © 1985.

6.4.3 Radio

- John H. Lienhard (1991). "Cox's Perpetual-Motion Machine". *The Engines of Our Ingenuity*. Episode 527http://www.uh.edu/engines/epi527.htm |transcripturl= missing title (help). NPR. KUHF-FM Houston.

- John H. Lienhard (1991). "John Joseph Merlin". *The Engines of Our Ingenuity*. Episode 630http://www.uh.edu/engines/epi630.htm |transcripturl= missing title (help). NPR. KUHF-FM Houston.

Chapter 7

Newman's energy machine

Newman's Energy Machine is a DC motor that its inventor Joseph Westley Newman claims produces mechanical power exceeding the electrical power being supplied to it (an over-unity or perpetual motion device). In 1979, Newman's patent application was rejected by the United States Patent Office, appealed, and rejected again, although various engineering experts supported Newman's claims amid controversy that the National Bureau of Standards failed to test the machine properly.

7.1 Claims by the inventor

By adding rolls to the armature of a motor, a larger and larger counter-electromotive force is generated on the motor. Newman outlined his claims about there being a fundamental electromagnetic interaction in all matter ultimately derived from only one type of force particle propagating at the speed of light. Newman claims that the motor derives its power by converting some of the mass of the copper in the coils into usable energy, in application of Einstein's Mass–energy equivalence. According to proponents of the Energy Machine, the most crucial part of the design concerns what happens as a result of mechanical commutation.[1][2]

7.2 U.S. patent application

In 1979, Newman submitted an application for his device to the United States Patent and Trademark Office.[2] The application was eventually rejected in 1983,[3] which set off a lengthy court battle. The United States District Court requested a master of the court to make the final decision. William E. Schuyler, Jr, former Commissioner of U.S. Patent Office, Washington, DC was chosen by the court to make the final decision to award the patent or not award the patent to Newman. Schuyler determined that evidence to support Newman's claim was overwhelming and found no contradictory factual evidence.[4]

In spite of more than thirty sworn affadavits from the scientific community supporting that Newman's machine worked as he claimed, as well as the former head of the patent office recommendation to award the patent, the judge ordered Newman's machine be confiscated and sent to, and be tested by, the National Bureau of Standards (NBS).

The National Bureau of Standards (NBS), now known as the National Institute of Standards and Technology (NIST), by request of the patent office, tested the device for several months and got negative results. In every case presented in the NBS report, the output power was less than power input from the battery pack, and therefore the efficiency was less than 100%. The court therefore upheld the rejection of the patent application.[5][6]

The NBS concluded in June 1986 that output power was not greater than the input, and it was not a perpetual motion machine. This report begins the controversy of Newman's Energy Machine because at the time of the NBS testing, over 30 independent scientists and engineers from around the world came the the laboratory of Joseph Newman and tested the machine for themselves, and presented their findings to the court in support of Newmans technology in the form of sworn affidavits. One such engineer, Dr. Roger Hastings, Chief Physicist of Unisys corporation at that time closely reviewed the testing of the NBS and declared: "Considering the importance of Newman's machine and its potential applications, this waste of NBS resources and misrepresentation of Newman's device is an insult to those seriously interested in the machine and to those who may benefit by its future applications", he continues, (speaking about the testers) "The primary r.f. energy generated by the machine was shunted to ground", and "Their measurements are therefore irrelevant to the actual functioning of the Newman device". Despite allegations of incompetence and poor testing methods, the NBS rejected Newman's claims, and the patent was again denied.

Newman argued that he had been mistreated by the patent office, and tried to have his motor legalized directly by the US Congress. He obtained a hearing in 30 July 1986 in front of several senators, but he was unsuccessful. During the hearing, Newman refused to have the machine tested by independent experts, and senator John Glenn pointed out that his supposedly-independent expert actually had a prior business relationship with him.

The case is now cited in the USPTO's Manual of Patent Examining Procedure as an example of an "inoperative" invention that can't have any utility, concretely as a perpetual motion machine.[7]

7.3 Perpetual motion controversy

See also: Mass–energy equivalence and First law of thermodynamics

Newman claims that his device derives its power by converting a small fraction of the mass in the copper coils into energy, and that it is therefore not a perpetual motion machine.[2] Many scientists don't think this theory is correct, and still classify it as "just another impossible perpetual motion machine". Skeptics argue that regardless of the exact mode of operation, if the output power is higher than the required input electrical power, the device should be capable of running "closed-loop", producing excess power without external batteries.

7.4 Legal controversy

In August 2007 the state of Alabama Securities Commission issued a cease and desist order against the "Newman Energy Corp", because it was selling unregistered securities of its company.[8]

7.5 See also

7.6 References

7.6.1 Bibliography

- Newman, J. (8th ed.).(1998). *The Energy Machine of Joseph Newman*. Scottsdale, AZ: Joseph Newman Publishing Company. 0-9613835-8-5

7.6.2 Notes

[1] "Perpetual Motion: Still Going Around". The Washington Post. 2000-01-12. Retrieved 2007-01-01.

[2] Peterson, Ivars, (1985-06-01). "A patent pursuit: Joe Newman's 'energy machine'.". Science News. Retrieved 2015-12-01.

[3] Newman, Joseph (1983-03-17). "Patent Application: "ENERGY GENERATION SYSTEM HAVING HIGHER ENERGY OUTPUT THAN INPUT" (failed)". Retrieved 2015-12-01.

[4] Kramer, Bruce (1988). "In Re Newman: The Federal Circuit Dismantles An Obstacle For Perpetual Motion Patent Applicants". *Akron Law Review*. Retrieved 2015-12-01.

[5] US Court of Appeals, Federal Circuit, Case #88-1312, *Newman v Quigg*.

[6] Peterson, Ivars, (5 July 1986). "NBS report short-circuits energy machine - National Bureau of Standards". Science News.

[7] *2107.01 General Principles Governing Utility Rejections (R-5) - 2100 Patentability. II. Wholly inoperative inventions; "incredible" utility*, U.S. Patent and Trademark Office Manual of Patent Examining Procedure

[8] Alabama Securities Commission (26 September 2008). "Administrative order C0-2007-0024 Consent order" (PDF).

7.7 External links

Claims

- "The Energy Machine of Joseph Newman" *(official site)*

- "The Newman's Energy Machine" by Jean-Louis Naudin and M. David

Skeptical

- "The Error Machine of Joseph Newman" by Tom Napier

- "Commentary: Crackpot Inventions" by James Randi

- "Is it possible to build a machine that generates more power than it uses?" by Cecil Adams

Chapter 8

Garabed T. K. Giragossian

slowly and put out a lot of energy for just a second.

8.1 See also

Voodoo Science, a book in which he is mentioned.

8.2 References

[1] *Automotive Industries* **39**, Chilton company, p. 34

- "*The Herald of Christ's Kingdom*" VOL. IX. August 1, 1926 No. 15

Garabed T. K. Giragossian was an Armenian living in Boston who is remembered for developing a perpetual motion device shortly after the turn of the 20th century. He immigrated to America in 1891. In 1917, Giragossian claimed, reportedly fraudulently, to have developed a "free energy machine". The assignment of the patent to the United States government was conditionally accepted in Pub. Res. 65-21, 40 Stat. 435, enacted on February 8, 1918, and authorized the Secretary of the Interior to form a committee to investigate the machine. The committee issued a report on July 1 finding that the principles were not sound.[1] Supposedly involved in a conspiracy, Woodrow Wilson signed a resolution offering him protection. The device was a giant flywheel that was charged up with energy

Chapter 9

Incredible utility

Main article: Utility (patent)

In United States patent law, **incredible utility** is a concept according to which, in order for an invention to be patentable, it must have some credible useful function. If it does not have a credible useful function despite the assertions of the inventor, then the application for patent can be rejected as having "incredible utility". The invention does not have to work the way the inventor thinks it works, but it must do something useful. Patents that have been held invalid for incredible utility include:

- an invention asserted to change the taste of food using a magnetic field (Fregeau v. Mossinghoff, 776 F.2d 1034, 227 USPQ 848 (Fed. Cir. 1985)),

- a perpetual motion machine (Newman v. Quigg, 877 F.2d 1575, 11 USPQ2d 1340 (Fed. Cir. 1989)),

- a flying machine operating on "flapping or flutter function" (In re Houghton, 433 F.2d 820, 167 USPQ 687 (CCPA 1970)),

- a cold fusion process for producing energy (In re Swartz, 232 F.3d 862, 56 USPQ2d 1703, (Fed. Cir. 2000)),

- a method for increasing the energy output of fossil fuels upon combustion through exposure to a magnetic field (In re Ruskin, 354 F.2d 395, 148 USPQ 221 (CCPA 1966)),

- uncharacterized compositions for curing a wide array of cancers (In re Citron, 325 F.2d 248, 139 USPQ 516 (CCPA 1963)), and

- a method of controlling the aging process (In re Eltgroth, 419 F.2d 918, 164 USPQ 221 (CCPA 1970)).[1]

A rejection based on incredible utility can be overcome by providing evidence that as a whole would lead a person having ordinary skill in the art to conclude that the asserted utility is more likely than not true.[2]

9.1 See also

- Industrial applicability
- Sufficiency of disclosure

9.2 References

[1] United States Patent and Trademark Office Manual of Patent Examination Procedures, 2107.01

[2] United States Patent and Trademark Office Manual of Patent Examination Procedures, 2107.02

Chapter 10

Infinite Energy (magazine)

Infinite Energy is a bi-monthly magazine published in New Hampshire that details theories and experiments concerning alternative energy, new science and new physics. The magazine was founded by the late Eugene Mallove, and is owned by the non-profit New Energy Foundation.[1] It was established in 1994 as *Cold Fusion* magazine [2] and changed its name in March 1995.[3]

Topics of interest include "new hydrogen physics," also called cold fusion; vacuum energy, or zero point energy; and so-called "environmental energy" which they define as the attempt to violate the Second Law of Thermodynamics,[4] for example with a perpetual motion machine. This is done in pursuit of the founder's commitment to "unearthing new sources of energy and new paradigms in science."[1] The magazine has also published articles and book reviews that are critical of the big bang theory that describes the origin of the universe.

The magazine has a print run of 3,000, and is available on U.S. newsstands. The issues range in size from 48 to 100 pages.

10.1 References

[1] Infinite Energy: What is the New Energy Foundation?

[2] Before 'Infinite Energy' it was 'Cold Fusion'

[3] About the magazine

[4] Infinite Energy FAQ: Environmental energy

10.2 External links

- Official website

Chapter 11

Magic wheel

The **magic wheel**, or **magnetic wheel** is a wheel that continues to spin for a long time after being started, and is one of the earliest examples of an attempt at a perpetual motion machine. This device was invented in medieval Bavaria. It looked like a wagon wheel spinning on an axle, affixed to a base. The superstitious population of the time believed it spun by the power of magic.

The mechanism of the magic wheel used several large magnets (lodestones) affixed to the wheel's outside rim, like the seats of a Ferris wheel. Each magnet was backed by a lead plate "seat". An extra stationary magnet was affixed to the base. Each magnet on the wheel's rim was attracted to the magnet in the base on its downward approach, then prevented from turning over when the opposite pole of the magnet passed over in the wheel, thus being repelled upward. The magnets were not allowed to touch one another. This attraction-repulsion maintained inertia efficiently, similar to a flywheel, such that the wheel spun for a very long time and was thought supernatural by some.

Incorrectly deemed by some to be a perpetual motion machine, the magic wheel eventually comes to a stop because of frictional losses at the central bearing. Proponents of free energy devices have advanced the theory that the lead plating interrupts the magnetic attraction between the rim magnets and the stationary magnet in sequence, thus permitting the wheel to continue turning and bring the next rim magnet into position. However, the presence of lead dampens magnetic fields equally in any directions (as a magnet's field lines must stay continuous from pole to pole), and the symmetry of closed forces in the system means that no interaction between rim magnets and the stationary magnets could generate the net increase in energy necessary to keep the wheel rotating.

The magic wheel was an impressive invention for the Dark Ages, a time when even some European kings were illiterate. An early German woodcut depicts a magic wheel.

11.1 See also

- History of perpetual motion machines

- Bhāskara's wheel

Chapter 12

Stefan Marinov

Stefan Marinov (Bulgarian: Стефан Маринов) (1 February 1931 – 15 July 1997) was a Bulgarian physicist, researcher, writer and lecturer who promoted anti-relativistic theoretical viewpoints, and later in his life defended the ideas of perpetual motion and free energy. In 1997 he self-published experimental results that confirmed classical electromagnetism and disproved that a machine constructed by Marinov himself could be a source of perpetual motion.[1][2] Devastated by the negative results he committed suicide[3] in Graz, Austria on 15 July 1997.

12.1   Life and education

Marinov was born on 1 February 1931 in Sofia to a family of intellectual communists.[4] In 1948 he finished Soviet College in Prague, then studied physics at the Czech Technical University in Prague and Sofia University. He was an Assistant Professor of Physics from 1960 to 1974 at Sofia University. In 1966-67, 1974, and 1977 he was subject to compulsory psychiatric treatment in Sofia because of his political dissent. In September 1977 Marinov received a passport and he successfully emigrated out of the country, moving to Brussels. In 1978, Marinov moved to Washington, D.C.. Later he lived in Italy and Austria. In his later years, Marinov earned a living as a groom for horses.

On 15 July 1997, Marinov jumped to his death from a staircase at a library at the University of Graz, after leaving suicide notes.[3] He was 66 years old and was survived by his son Marin Marinov, who at the time was a vice-Minister of Industry of Bulgaria.

12.2   Work

One of Marinov's interests was the quest for free energy sources via construction of toy theories (new axiomatic systems that putatively describe our physical reality) and their experimental testing against mainstream physical theories. In 1992 Marinov wrote a letter to German Federal Chan-cellor Helmut Kohl in support of a German company, Becocraft, that was doing research into "free energy" technologies and had recently been the target of lawsuits. In the letter, Marinov threatened to set himself on fire at the steps of the German parliament if Kohl was not willing to intervene in favour of Marinov's associates.

12.3   Research

Marinov attempted to find experimental disproof of the theory of relativity by testing the speed of light in different directions using an arrangement of *coupled mirrors* and *coupled shutters*.[5][6] Marinov reported in 1974 that he had measured an anisotropy of the velocity of light.[7] However, Marinov's claims have not found acceptance within the scientific community, despite his energetic efforts to promote his claims. Marinov planned to develop an updating of the relativistic mechanics and electrodynamics, as described in his self-published book *Eppur si Muove*.[4] Marinov succeeded in having his claims presented in numerous publications including peer-reviewed journals.[8][9][10][11][12][13][14][15][16][17][18][19][20][21]

Marinov was involved publicly with many quarrels with John Maddox, the Editor of Nature, who refused to print either his papers or his Letters to the Editor. He retaliated by securing the funds to place full-page advertisement in Nature expressing his frustration with what he regarded as the dogmatic attitude of the Establishment.[22] Marinov himself published a journal, *Deutsche Physik,* where he was Editor-In-Chief, and which discussed mainly his ideas on physics.

Stefan Marinov was interested in bizarre experiments alleged to violate known physical laws.[23][24] Marinov claimed to have seen in operation and learned the secret of the so-called "Swiss ML converter" or Testatika electrical generator, another alleged perpetual motion machine, at a religious commune in Switzerland called Methernitha.[3] According to Marinov's account, this 500-member commune, led by religious leader Paul Baumann, met all its energy needs using this device.[3]

Marinov has been editor of a 5 volume encyclopaedic series called "Classical Physics".[25][26][27][28][29] In 1993 Marinov also authored a book on electromagnetism[30] which discoursed on his belief that mainstream scientific thought was mired in dogma, and had discarded still-valid knowledge from scientific thought of previous eras. In 1997 in the last issue 21 of *Deutsche Physik* Marinov self-published experimental results that disprove that the *Siberian Coliu*, constructed by Marinov himself, is a perpetual motion machine, and where Marinov concluded that Ampere's law in electromagnetism is correct.[1][2] Most of Marinov's friends think these negative results on constructing a source of free energy (in order to solve the global energetic needs of humanity) might have pushed Marinov to commit a suicide.[3][31]

12.4 References

[1] Marinov S (1997). "Siberian Coliu machine with eccentric circular current rotor". *Deutsche Physik* **6** (21): 5–36. External link in |journal= (help)

[2] Marinov S (1997). "Editor's comments on "A history of the theories of aether and electricity by E. Whittaker"". *Deutsche Physik* **6** (21): 56. External link in |journal= (help)

[3] Schneeberger E, Bass R (1997). "Stefan Marinov: In Memoriam: My Scientific Testament; A Strong Voice Is Missing (Last Will and Testament); Letter from Erwin Schneeberger; Letter from Dr. Robert W. Bass". *New Energy News* **5** (5): 1–3.

[4] Marinov S (1987). "Eppur Si Muove: Axiomatics, fundamentals and experimental verification of the absolute space-time theory". *East-West Publishers, Graz, Austria.*

[5] Stefan Marinov (1983). "The interrupted 'rotating disc' experiment" (PDF). *Journal of Physics A* **16**: 1885–1888. Bibcode:1983JPhA...16.1885M. doi:10.1088/0305-4470/16/9/013.

[6] Marinov S (2007). "New Measurement of the Earth's Absolute Velocity with the Help of the Coupled Shutters Experiment" (PDF). *Progress in Physics* **1**: 31–37.

[7] Marinov S (1974). "The velocity of light is direction dependent". *Czechoslovak Journal of Physics B* **24** (9): 965–970. Bibcode:1974CzJPh..24..965M. doi:10.1007/BF01591047.

[8] Marinov S (1972). "How to measure the earth's velocity with respect to absolute space". *Physics Letters A* **41** (5): 433–434. Bibcode:1972PhLA...41..433M. doi:10.1016/0375-9601(72)90392-1.

[9] Marinov S (1970). "Experimentum crucis for the proof of the space-time absoluteness". *Physics Letters A* **32** (3): 183–184. Bibcode:1970PhLA...32..183M. doi:10.1016/0375-9601(70)90265-3.

[10] Marinov S (1972). "Concerning the experimentum crucis for the proof of the space-time absoluteness". *Physics Letters A* **40** (1): 73–74. Bibcode:1972PhLA...40...73M. doi:10.1016/0375-9601(72)90205-8.

[11] Marinov S (1973). "Kantor's second-order Doppler-effect experiment treated by the absolute space-time theory". *Physics Letters A* **44** (1): 21–22. Bibcode:1973PhLA...44...21M. doi:10.1016/0375-9601(73)90941-9.

[12] Marinov S (1974). "Velocity of light in a moving medium according to the absolute space-time theory". *International Journal of Theoretical Physics* **9** (2): 139–144. Bibcode:1974IJTP....9..139M. doi:10.1007/BF01807696.

[13] Marinov S (1975). "A reliable experiment for the proof of the space-time absoluteness". *Physics Letters A* **54** (1): 19–20. Bibcode:1975PhLA...54...19M. doi:10.1016/0375-9601(75)90589-7.

[14] Marinov S (1976). *New Scientist* **71**: 662. Missing or empty |title= (help)

[15] Marinov S (1976). "Gravitational (dynamic) time dilation according to absolute space-time theory". *Foundations of Physics* **6** (5): 571–581. Bibcode:1976FoPh....6..571M. doi:10.1007/BF00715109.

[16] Marinov S (1977). "A pure experiment to establish that the velocity of light does not depend on the velocity of the source". *Physics Letters A* **62** (5): 293–294. Bibcode:1977PhLA...62..293M. doi:10.1016/0375-9601(77)90419-4.

[17] Marinov S (1978). "Rotating disk experiments". *Foundations of Physics* **8** (1-2): 137–156. Bibcode:1978FoPh....8..137M. doi:10.1007/BF00708494.

[18] Marinov S (1978). "The light Doppler effect treated by absolute spacetime theory". *Foundations of Physics* **8** (7-8): 637–652. Bibcode:1978FoPh....8..637M. doi:10.1007/BF00717587.

[19] Marinov S (1979). "The coordinate transformations of the absolute space-time theory". *Foundations of Physics* **9** (5-6): 445–460. Bibcode:1979FoPh....9..445M. doi:10.1007/BF00708535.

[20] Marinov S (1980). "Measurement of the laboratory's absolute velocity" (PDF). *General Relativity and Gravitation* **12** (1): 57–66. Bibcode:1980GReGr..12...57M. doi:10.1007/BF00756168.

[21] Marinov S (1982). "Measurement of the one-way speed of light and the Earth's absolute velocity". *Proceeding of 2nd Marcel Grossmann Meeting, Trieste, Italy*: 547–550.

[22] Stefan Marinov (1996). "Annus Horribilis". *Nature* **380** (6572): xiv.

[23] Stefan Marinov (1989). "The Intriguing ball-bearing motor". *Electronics World & Wireless World* (Reed Business Publishing) (April).

[24] Frank Ogden (1989). "Great Balls of Fire!". *Electronics World & Wireless World* (Reed Business Publishing) (April).

[25] Marinov S (1981). "Classical Physics: Part I,". *East-West Publishers, Graz, Austria.*

[26] Marinov S (1981). "Classical Physics: Part II,". *East-West Publishers, Graz, Austria.*

[27] Marinov S (1981). "Classical Physics: Part III, High-Velocity Mechanics". *East-West Publishers, Graz, Austria.*

[28] Marinov S (1981). "Classical Physics: Part IV,". *East-West Publishers, Graz, Austria.*

[29] Marinov S (1981). "Classical Physics: Part V, Electromagnetism". *East-West Publishers, Graz, Austria.*

[30] Marinov S (1993). "Divine Electromagnetism". *East West Publishers, Graz.*

[31] Pappas P (1997). "Update on Stefan Marinov's Death (E-mails by Panos Pappas)".

Chapter 13

Maxwell's demon

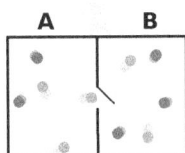 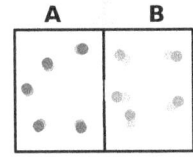

Schematic figure of Maxwell's demon

In the philosophy of thermal and statistical physics, **Maxwell's demon** is a thought experiment created by the physicist James Clerk Maxwell in which he suggested how the Second Law of Thermodynamics could hypothetically be violated.[1] In the thought experiment, a demon controls a small door between two chambers of gas. As individual gas molecules approach the door, the demon quickly opens and shuts the door so that slow molecules pass into one chamber and fast molecules pass into the other chamber. Because faster molecules are hotter, the demon's behavior causes one chamber to warm up as the other cools, thus decreasing entropy and violating the Second Law of Thermodynamics.

13.1   Origin and history of the idea

The thought experiment first appeared in a letter Maxwell wrote to Peter Guthrie Tait on 11 December 1867. It appeared again in a letter to John William Strutt in 1871, before it was presented to the public in Maxwell's 1872 book on thermodynamics titled *Theory of Heat*.[2]

In his letters and books, Maxwell described the agent opening the door between the chambers as a "finite being". William Thomson (Lord Kelvin) was the first to use the word "demon" for Maxwell's concept, in the journal *Nature* in 1874, and implied that he intended the mediating, rather than malevolent, connotation of the word.[3][4][5]

13.2   Original thought experiment

The second law of thermodynamics ensures (through statistical probability) that two bodies of different temperature, when brought into contact with each other and isolated from the rest of the Universe, will evolve to a thermodynamic equilibrium in which both bodies have approximately the same temperature.[6] The second law is also expressed as the assertion that in an isolated system, entropy never decreases.[6]

Maxwell conceived a thought experiment as a way of furthering the understanding of the second law. His description of the experiment is as follows:[6][7]

> ... if we conceive of a being whose faculties are so sharpened that he can follow every molecule in its course, such a being, whose attributes are as essentially finite as our own, would be able to do what is impossible to us. For we have seen that molecules in a vessel full of air at uniform temperature are moving with velocities by no means uniform, though the mean velocity of any great number of them, arbitrarily selected, is almost exactly uniform. Now let us suppose that such a vessel is divided into two portions, A and B, by a division in which there is a small hole, and that a being, who can see the individual molecules, opens and closes this hole, so as to allow only the swifter molecules to pass from A to B, and only the slower molecules to pass from B to A. He will thus, without expenditure of work, raise the temperature of B and lower that of A, in contradiction to the second law of thermodynamics.

In other words, Maxwell imagines one container divided into two parts, *A* and *B*.[8][6] Both parts are filled with the same gas at equal temperatures and placed next to each other. Observing the molecules on both sides, an imaginary demon guards a trapdoor between the two parts. When a

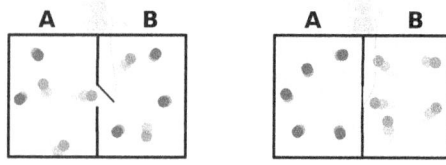

Schematic figure of Maxwell's demon

faster-than-average molecule from *A* flies towards the trapdoor, the demon opens it, and the molecule will fly from *A* to *B*. Likewise, when a slower-than-average molecule from *B* flies towards the trapdoor, the demon will let it pass from *B* to *A*. The average speed of the molecules in *B* will have increased while in *A* they will have slowed down on average. Since average molecular speed corresponds to temperature, the temperature decreases in *A* and increases in *B*, contrary to the second law of thermodynamics. A heat engine operating between the thermal reservoirs *A* and *B* could extract useful work from this temperature difference.

The demon must allow molecules to pass in both directions in order to produce only a temperature difference; one-way passage only of faster-than-average molecules from *A* to *B* will cause higher temperature *and* pressure to develop on the *B* side.

13.3   Criticism and development

Several physicists have presented calculations that show that the second law of thermodynamics will not actually be violated, if a more complete analysis is made of the whole system including the demon.[8][6][9] The essence of the physical argument is to show, by calculation, that any demon must "generate" more entropy segregating the molecules than it could ever eliminate by the method described. That is, it would take more thermodynamic work to gauge the speed of the molecules and selectively allow them to pass through the opening between *A* and *B* than the amount of energy gained by the difference of temperature caused by this.

One of the most famous responses to this question was suggested in 1929 by Leó Szilárd,[10] and later by Léon Brillouin.[8][6] Szilárd pointed out that a real-life Maxwell's demon would need to have some means of measuring molecular speed, and that the act of acquiring information would require an expenditure of energy. Since the demon and the gas are interacting, we must consider the total entropy of the gas and the demon combined. The expenditure of energy by the demon will cause an increase in the entropy of the demon, which will be larger than the lowering of the entropy of the gas.

In 1960, Rolf Landauer raised an exception to this argument.[11][8][6] He realized that some measuring processes need not increase thermodynamic entropy as long as they were thermodynamically reversible. He suggested these "reversible" measurements could be used to sort the molecules, violating the Second Law. However, due to the connection between thermodynamic entropy and information entropy, this also meant that the recorded measurement must not be erased. In other words, to determine whether to let a molecule through, the demon must acquire information about the state of the molecule and either discard it or store it. Discarding it leads to immediate increase in entropy but the demon cannot store it indefinitely: In 1982, Charles Bennett showed that, however well prepared, eventually the demon will run out of information storage space and must begin to erase the information it has previously gathered.[8][12] Erasing information is a thermodynamically irreversible process that increases the entropy of a system. Although Bennett had reached the same conclusion as Szilard's 1929 paper, that a Maxwellian demon could not violate the second law because entropy would be created, he had reached it for different reasons. Regarding Landauer's principle, the minimum energy dissipated by deleting information was experimentally measured by Eric Lutz *et al.* in 2012.[13]

John Earman and John D. Norton have argued that Szilárd and Landauer's explanations of Maxwell's demon begin by assuming that the second law of thermodynamics cannot be violated by the demon, and derive further properties of the demon from this assumption, including the necessity of consuming energy when erasing information, etc.[14][15] It would therefore be circular to invoke these derived properties to defend the second law from the demonic argument. Bennett later acknowledged the validity of Earman and Norton's argument, while maintaining that Landauer's principle explains the mechanism by which real systems do not violate the second law of thermodynamics.[16]

13.4   Recent progress

Although the argument by Landauer and Bennett only answers the consistency between the second law of thermodynamics and the whole cyclic process of the entire system of Szilard engine (a composite system of the engine and the demon), recent approach based on the non-equilibrium thermodynamics for small fluctuating systems has provided deeper insight on each information process with each subsystem. From this viewpoint, the measurement process is regarded as a process where the correlation (mutual information) between the engine and the demon increases, and the feedback process is regarded as a process where the correlation decreases. If the correlation changes, thermodynamic relations as the second law of thermodynamics and the fluctuation theorem for each subsystem should be modified, and for the case of external control a second-law like inequality[17] and a generalized fluctuation theorem[18] with mutual information are satisfied. These relations suggest that we need extra thermodynamic cost to increase correlation (measurement case), and in contrast we can apparently violate the second law up to the consumption of correlation (feedback case). For more general information processes including biological information processing, both inequality[19] and equality[20] with mutual information hold.

13.5 Applications

Real-life versions of Maxwellian demons occur, but all such "real demons" have their entropy-lowering effects duly balanced by increase of entropy elsewhere. Molecular-sized mechanisms are no longer found only in biology; they are also the subject of the emerging field of nanotechnology. Single-atom traps used by particle physicists allow an experimenter to control the state of individual quanta in a way similar to Maxwell's demon.

If hypothetical mirror matter exists, Zurab Silagadze proposes that demons can be envisaged, "which can act like perpetuum mobiles of the second kind: extract heat energy from only one reservoir, use it to do work and be isolated from the rest of ordinary world. Yet the Second Law is not violated because the demons pay their entropy cost in the hidden (mirror) sector of the world by emitting mirror photons."[21]

13.6 Experimental work

In the February 2007 issue of *Nature*, David Leigh, a professor at the University of Edinburgh, announced the creation of a nano-device based on the Feynman's Brownian ratchet. Leigh's device is able to drive a chemical system out of equilibrium, but it must be powered by an external source (light in this case) and therefore does not violate thermodynamics.

Previously, other researchers created a ring-shaped molecule which could be placed on an axle connecting two sites, A and B. Particles from either site would bump into the ring and move it from end to end. If a large collection of these devices were placed in a system, half of the devices had the ring at site A and half at B, at any given moment in time.

Leigh made a minor change to the axle so that if a light is shone on the device, the center of the axle will thicken, restricting the motion of the ring. It only keeps the ring

from moving, however, if it is at A. Over time, therefore, the rings will be bumped from B to A and get stuck there, creating an imbalance in the system. In his experiments, Leigh was able to take a pot of "billions of these devices" from 50:50 equilibrium to a 70:30 imbalance within a few minutes.[22]

In 2009 Mark G. Raizen developed a laser atomic cooling technique which realizes the process Maxwell envisioned of sorting individual atoms in a gas into different containers based on their energy.[6][23][24] The new concept is a one-way wall for atoms or molecules that allows them to move in one direction, but not go back. The operation of the one-way wall relies on an irreversible atomic and molecular process of absorption of a photon at a specific wavelength, followed by spontaneous emission to a different internal state. The irreversible process is coupled to a conservative force created by magnetic fields and/or light. Raizen and collaborators proposed to use the one-way wall in order to reduce the entropy of an ensemble of atoms. In parallel, Gonzalo Muga and Andreas Ruschhaupt, independently developed a similar concept. Their "atom diode" was not proposed for cooling, but rather to regulate flow of atoms. The Raizen Group demonstrated significant cooling of atoms with the one-way wall in a series of experiments in 2008. Subsequently, the operation of a one-way wall for atoms was demonstrated by Daniel Steck and collaborators later in 2008. Their experiment was based on the 2005 scheme for the one-way wall, and was not used for cooling. The cooling method realized by the Raizen Group was called "Single-Photon Cooling," because only one photon on average is required in order to bring an atom to near-rest. This is in contrast to other laser cooling techniques which uses the momentum of the photon and requires a two-level cycling transition.

In 2006 Raizen, Muga, and Ruschhaupt showed in a theoretical paper that as each atom crosses the one-way wall, it scatters one photon, and information is provided about the turning point and hence the energy of that particle. The entropy increase of the radiation field scattered from a directional laser into a random direction is exactly balanced by the entropy reduction of the atoms as they are trapped with the one-way wall.

This technique is widely described as a "Maxwell's demon" because it realizes Maxwell's process of creating a temperature difference by sorting high and low energy atoms into different containers. However scientists have pointed out that it is not a true Maxwell's demon in the sense that it does not violate the second law of thermodynamics;[25][6] it does not result in a net decrease in entropy[25][6] and cannot be used to produce useful energy. This is because the process requires more energy from the laser beams than could be produced by the temperature difference generated. The atoms absorb low entropy photons from the laser beam and emit them in a random direction, thus increasing the entropy of the environment.[25][6]

13.7   As metaphor

Historian Henry Brooks Adams in his manuscript *The Rule of Phase Applied to History* attempted to use Maxwell's demon as a historical metaphor, though he misunderstood and misapplied the original principle.[26] Adams interpreted history as a process moving towards "equilibrium", but he saw militaristic nations (he felt Germany pre-eminent in this class) as tending to reverse this process, a Maxwell's demon of history. Adams made many attempts to respond to the criticism of his formulation from his scientific colleagues, but the work remained incomplete at Adams' death in 1918. It was only published posthumously.[27]

Sociologist Pierre Bourdieu incorporated Maxwell's demon into his work, "Raisons Pratiques" as a metaphor for the socioeconomic inequality among students, as maintained by the school system, the economy, and families.

The demon is mentioned several times in The Cyberiad, a series of short stories by the noted science fiction writer Stanisław Lem. In the book the demon appears both in its original form and in a modified form where it uses its knowledge of all particles in the box in order to surmise general (but unfocused and random) facts about the rest of the universe.

A machine powered by Maxwell's demon plays a role in Thomas Pynchon's novel The Crying of Lot 49. Maxwells Demon play the role of spirit of perversity in Christopher Stasheffs books Her Majesty's Wizard and the Witch Doctor with the ability to concentrate or diffuse heat/energy

13.8   See also

- Brownian ratchet
- *Chance and Necessity*
- Catalysis
- Dispersive mass transfer
- Evaporation
- Gibbs paradox
- Hall effect
- Heisenberg's Uncertainty Principle
- Joule–Thomson effect

- Laplace's demon
- Laws of thermodynamics
- Mass spectrometry
- *Maxwell's Maniac*
- Photoelectric effect
- Quantum tunnelling
- Schrödinger's cat
- Thermionic emission
- Vortex tube

13.9   Notes

[1] Cargill Gilston Knott (1911). "Quote from undated letter from Maxwell to Tait". *Life and Scientific Work of Peter Guthrie Tait*. Cambridge University Press. p. 215.

[2] Leff & Rex (2002), p. 370.

[3] William Thomson (1874). "Kinetic theory of the dissipation of energy". *Nature* **9** (232): 441–444. Bibcode:1874Natur...9..441T. doi:10.1038/009441c0.

[4] "The sorting demon Of Maxwell". *Proceedings of the Royal Institution* **ix**: 113. 1879.

[5] Alan S. Weber (2000). *Nineteenth Century Science: a Selection of Original Texts*. Broadview Press. p. 300.

[6] Bennett, Charles H. (November 1987). "Demons, Engines, and the Second Law" (PDF). *Scientific American* (Scientific American Inc.) **257** (5): 108–116. doi:10.1038/scientificamerican1187-108. Retrieved November 13, 2014.

[7] Maxwell (1871), reprinted in Leff & Rex (1990) on p. 4.

[8] Sagawa, Takahiro (2012). *Thermodynamics of Information Processing in Small Systems*. Springer Science and Business Media. pp. 9–14. ISBN 4431541675.

[9] Bennett, Charles H.; Schumacher, Benjamin (August 2011). "Maxwell's demons appear in the lab" (PDF). *Nikkei Science* (Scientific American Inc.): 3–6. Retrieved November 13, 2014.

[10] Szilard, Leo (1929). "Über die Entropieverminderung in einem thermodynamischen System bei Eingriffen intelligenter Wesen (On the reduction of entropy in a thermodynamic system by the intervention of intelligent beings)". *Zeitschrift für Physik* **53**: 840–856. Bibcode:1929ZPhy...53..840S. doi:10.1007/bf01341281. cited in Bennett 1987. English translation available as NASA document TT F-16723 published 1976

[11] Landauer, R. (1961). "Irreversibility and heat generation in the computing process" (PDF). *IBM Jour. of Research and Development* (International Business Machines) **5** (3): 183–191. doi:10.1147/rd.53.0183. Retrieved November 13, 2014. reprinted in Vol. 44, No. 1, January 2000, p. 261

[12] Bennett, C. H. (1982). "The thermodynamics of computation—a review". *International Journal of Theoretical Physics* **21** (12): 905–940. doi:10.1007/BF02084158.

[13] jobs (2012-03-07). "The unavoidable cost of computation revealed : Nature News & Comment". Nature.com. Retrieved 2012-09-07.

[14] John Earman and John D. Norton (1998). "Exorcist XIV: The Wrath of Maxwell's Demon. Part I. From Maxwell to Szilard" (PDF). *Studies in the History and Philosophy of Modern Physics* **29**: 435. doi:10.1016/s1355-2198(98)00023-9.

[15] John Earman and John D. Norton (1999). "Exorcist XIV: The Wrath of Maxwell's Demon. Part II. From Szilard to Landauer and Beyond" (PDF). *Studies in the History and Philosophy of Modern Physics* **30**: 1. doi:10.1016/s1355-2198(98)00026-4.

[16] Charles H. Bennett (2002–2003). "Notes on Landauer's principle, reversible computation, and Maxwell's demon" (PDF). arXiv:physics/0210005. Bibcode:2002physics..10005B.

[17] Hugo Touchette and Seth Lloyd (2000). "Information-Theoretic Limits of Control". *Physical Review Letters* **84**: 1156–1159. Bibcode:2000PhRvL..84.1156T. doi:10.1103/PhysRevLett.84.1156.

[18] Takahiro Sagawa and Masahito Ueda (2010). "Generalized Jarzynski Equality under Nonequilibrium Feedback Control". *Physical Review Letters* **104**: 090602. arXiv:0907.4914. Bibcode:2010PhRvL.104i0602S. doi:10.1103/PhysRevLett.104.090602.

[19] Armen E Allahverdyan, Dominik Janzing and Guenter Mahler (2009). "Thermodynamic efficiency of information and heat flow". *Journal of Statistical Mechanics* **2009**: P09011. arXiv:0907.3320. Bibcode:2009JSMTE..09..011A. doi:10.1088/1742-5468/2009/09/P09011.

[20] Naoto Shiraishi and Takahiro Sagawa (2015). "Fluctuation theorem for partially masked nonequilibrium dynamics". *Physical Review E* **91**: 012130. arXiv:1403.4018. Bibcode:2015PhRvE..91a2130S. doi:10.1103/PhysRevE.91.012130.

[21] "[physics/0608114] Maxwell's demon through the looking glass". Uk.arxiv.org. 2006-08-10. Retrieved 2013-02-18.

[22] Katharine Sanderson (31 January 2007). "A demon of a device". Nature. doi:10.1038/news070129-10.

[23] Raizen, Mark G. (June 12, 2009). "Comprehensive Control of Atomic Motion". *Science* (American Assoc. for the Advancement of Science) **324** (5933): 1403–1406. Bibcode:2009Sci...324.1403R. doi:10.1126/science.1171506. Retrieved November 14, 2014.

[24] Raizen, Mark G. (March 2011). "Demons, Entropy, and the Quest for Absolute Zero". *Scientific American* (Scientific American Inc.) **304** (3): 54–59. doi:10.1038/scientificamerican0311-54. Retrieved November 14, 2014.

[25] Orzel, Chad (January 25, 2010). "Single-Photon Cooling: Making Maxwell's Demon". *Uncertain Principles*. ScienceBlogs website. Retrieved November 14, 2014.

[26] Cater (1947), pp. 640–647; see also Daub (1970), reprinted in Leff & Rex (1990), pp. 37–51.

[27] Adams (1919), p. 267.

13.10   References

- Cater, H. D., ed. (1947). *Henry Adams and his Friends*. Boston.

- Daub, E. E. (1967). "Atomism and Thermodynamics". *Isis* **58** (3): 293–303. doi:10.1086/350264.

- Leff, Harvey S. & Andrew F. Rex, ed. (1990). *Maxwell's Demon: Entropy, Information, Computing*. Bristol: Adam-Hilger. ISBN 0-7503-0057-4.

- Leff, Harvey S. & Andrew F. Rex, ed. (2002). *Maxwell's Demon 2: Entropy, Classical and Quantum Information, Computing*. CRC Press. ISBN 0-7503-0759-5.

- Adams, H. (1919). *The Degradation of the Democratic Dogma*. New York: Kessinger. ISBN 1-4179-1598-6.

- Vladislav Cápek & Daniel P. Sheehan (2005). *Challenges to the Second Law of Thermodynamics*. The Netherlands: Springer. ISBN 1-4020-3016-9.

13.11   External links

- Bennett, C. H. (1987) "Demons, Engines and the Second Law", *Scientific American*, November, *pp*108-116

- Binder, P.-M. (2008). "Reflections on a Wall of Light". *Science* **322** (5906): 1334–1335. doi:10.1126/science.1166681.

- Earman, J. and Norton, J. (1998). "Exorcist XIV: The Wrath of Maxwell's Demon. Part I. From Maxwell to Szilard" (PDF). *Studies in History and Philosophy of Science Part B: Studies in History and Philosophy of Modern Physics* **29** (4): 435–471. doi:10.1016/S1355-2198(98)00023-9.

- Earman, J. and Norton, J. (1999). "Exorcist XIV: The Wrath of Maxwell's Demon. Part II. From Szilard to Landauer and Beyond" (PDF). *Studies in History and Philosophy of Science Part B: Studies in History and Philosophy of Modern Physics* **30**: 1–40. doi:10.1016/S1355-2198(98).

- Feynman, R. P.; et al. (1996). *Feynman Lectures on Computation*. Addison-Wesley. pp. 148–150. ISBN 0-14-028451-6.

- Jordy, W. H. (1952). *Henry Adams: Scientific Historian*. New Haven. ISBN 0-685-26683-4.

- Khan, Salman. "Maxwell's Demon".

- Maroney, O. J. E. (2009) ""Information Processing and Thermodynamic Entropy" The Stanford Encyclopedia of Philosophy (Autumn 2009 Edition)

- Maxwell, J. C. (1871). *Theory of Heat.*, reprinted (2001) New York: Dover, ISBN 0-486-41735-2

- Norton, J. (2005). "Eaters of the lotus: Landauer's principle and the return of Maxwell's demon" (PDF). *Studies in History and Philosophy of Science Part B: Studies in History and Philosophy of Modern Physics* **36** (2): 375–411. doi:10.1016/j.shpsb.2004.12.002.

- Raizen, Mark G. (2011) "Demons, Entropy, and the Quest for Absolute Zero", *Scientific American,* March, *pp*54-59

- Reaney, Patricia. "Scientists build nanomachine", *Reuters*, February 1, 2007

- Rubi, J Miguel, "Does Nature Break the Second Law of Thermodynamics?"; Scientific American, October 2008 :

- Splasho (2008) - Historical development of Maxwell's demon

- Weiss, Peter. "Breaking the Law - Can quantum mechanics + thermodynamics = perpetual motion?", *Science News*, October 7, 2000

Chapter 14

Methernitha

Methernitha refers to two related entities, both founded by Paul Baumann — **Methernitha Christian Alliance** and **Methernitha Cooperative**. One is a religious group, and the other is a community in Linden, Switzerland, based on the group's principles.

14.1 Organizations

14.1.1 The religious group

The Methernitha Christian Alliance exists as a loose association of members who subscribe to the teachings of the Holy Scriptures. Members do not have organized meetings, nor do they necessarily live in the same area. Current members are distributed throughout Switzerland and other countries. The Alliance was founded in the 1950s by a small group of Christian oriented people whose goal was to live according to the teachings of the Bible. Eventually the group founded a commune in Linden in the canton of Bern.

14.1.2 The community

Originally based on the principles of the Methernitha Christian Alliance, the Methernitha Cooperative is now more philosophically inclusive and in 1960 was registered with the Swiss Commercial Register as a legal entity. The organisation was laid down in Articles of Association and adheres to the Rules of the Swiss Federation.

The Cooperative is a residential joint venture work group operated on a democratic basis. Administration staff are chosen from the members in a general meeting. It has its own television studios that transmit self-produced programs. The commune also has its own school and nursing home, and a "club house", which is the center for worship and their historical archives.

Members are still Swiss citizens, and the Cooperative pays all relevant taxes and social security to the government.

14.2 Beliefs

Methernitha is a Christian group, but it is non-evangelical, refraining from attempts to win new converts. Members do not indulge in alcohol, tobacco or recreational drugs or money.

14.3 Free energy

The Methernitha Cooperative claimed to have developed years ago a free-energy[1] device called the Testatika:

> "This wonder-machine is fueled from nature, nothing else. Nature is the greatest source of power as well as knowledge which man has, and it still conceals many secrets, which are only revealed to those, who approach and tie in with them with highest respect and responsibility ... To understand nature and to perceive its voice, man is obliged to experience silence and solitude, and it was there, where the knowledge about this technology was obtained."

14.4 Criticism

In France, the religious group was criticized by an anti-cult association ADFI, which considered it a cult ("secte" in French-language).[2] In 1976 Paul Baumann, the founder of the community, was sentenced to six years imprisonment for sexual abuse of children.[3]

14.5 References

[1] Carolina Hehenkamp (2 February 2002). "Methernitha — A Community That Runs on Free Energy and Spiritual Values". *Spiritofmaat*. Retrieved 27 August 2009.

[2] UNADFI (1992). "Association spirituelle ou entreprise plutôt lucrative ?" (in French). *Prevensectes*. Retrieved 27 August 2009.

[3] Martin Baumann, Jörg Stolz (2007). *Eine Schweiz, viele Religionen: Risiken und Chancen des Zusammenlebens* (in German). p. 295. Retrieved 27 August 2009.

14.6 External links

- Official website

Chapter 15

Perepiteia

For a term developed by Aristotle in his work Poetics, see peripeteia.

Perepiteia is claimed to be a new generator developed by the Canadian inventor Thane Heins. The device is named after the Greek word for peripety, a dramatic reversal of circumstances or turning point in a story. The device was quickly attributed the term "perpetual motion machine" by several media outlets. Due to the long history of hoaxes and failures of perpetual motion machines and the incompatibility of such a device with accepted principles of physics, Heins' claims about Perepiteia have been treated with considerable skepticism.

In 2003, Heins filed a patent application in Canada[1] but no patent was granted.[2] Heins also founded Potential Difference Inc, the website of which contains a series of videos of the inventor demonstrating the machine.[3]

Heins has recently stated that he is unsure whether or not the machine really produces energy, but in communications with science writer David Bradley of ScienceBase, Heins made claims of up to 7000% efficiency for a bi-toroidal transformer.[4] Heins, who reportedly works 8–12 hours a day on the Perepiteia, insists that it is viable and that "This technology should be mainstream."[5]

15.1 Theory

Mechanically, the device appears to be an induction motor with a magnetic material placed inside the rotor core.[6] Heins believes that the device's potential may rest in its atypical manipulation of the back electromotive force (back EMF). A more detailed description of the device may be found in the patent application, minus supporting figures.[1]

The apparent unique quality of the Perepiteia machine is that instead of maintaining a certain state of motion, it appears to generate acceleration. According to Heins, the Perepiteia produces magnetic friction which somehow gets turned into a magnetic boost. Using an electric motor, the drive shaft is attached to a steel rotor with small round magnets lining its outer edges. In this set-up of a simple generator, the rotor spins so that the magnets pass by a wire coil just in front of them, generating electrical energy.

15.2 Operation

Perepiteia's process begins by overloading the generator to get a current, which typically causes the wire coil to build up a large electromagnetic field. Usually, this kind of electromagnetic field creates an effect called the back electromotive force (back EMF) due to Lenz's law. The effect should repel the spinning magnets on the rotor, and slow them down until the motor stops completely, in accordance with the law of conservation. However, instead of stopping, the rotor accelerates - i.e. the magnetic friction did not repel the magnets and wire coil. Heins states that the steel rotor and driveshaft had conducted the magnetic resistance away from the coil and back into the electric motor. In effect, the back EMF was boosting the magnetic fields used by the motor to generate electrical energy and cause acceleration. The faster the motor accelerated, the stronger the electromagnetic field it would create on the wire coil, which in turn would make the motor go even faster. Heins seemed to have created a positive feedback loop. To confirm the theory, Heins replaced part of the driveshaft with plastic pipe that wouldn't conduct the magnetic field. There was no acceleration.[7]

15.3 Scientific examination

In early 2008, Heins was given access to equipment to demonstrate it by professor Riadh Habash of the University of Ottawa, who says of it, "It accelerates, but when it comes to an explanation, there is no backing theory for it. That's why we're consulting MIT. But at this time we can't support any claim."[8]

After examining the machine and witnessing a demonstration, Massachusetts Institute of Technology (MIT) professor Markus Zahn admitted that he could not fully explain its operation. Although he refused to call it perpetual motion, he stated that it might be an extremely efficient motor.[9] Regarding the device, Zahn stated that "It's an unusual phenomena [*sic*] I wouldn't have predicted in advance. But I saw it. It's real. Now I'm just trying to figure it out...To my mind this is unexpected and new, and it's worth exploring all the possible advantages once you're convinced it's a real effect."[10] However, even if Perepiteia does not produce perpetual motion, Zahn still believes that the device could have considerable practical applications, noting that "There are an infinite number of induction machines in people's homes and everywhere around the world. If you could make them more efficient, cumulatively, it could make a big difference."[7]

However, Zahn later stated in an interview that "I can't understand how [Heins] can even breathe the words 'perpetual motion.' He plugs it into the wall." In a subsequent e-mail to Heins, Zahn wrote that: "Any talk of perpetual motion, over unity efficiency, etc. discredits you, now me, and your ideas." Zahn further stated that he would not endorse Heins' device until "the foolishness is stopped of hinting that your motor violates fundamental laws of physics."[11]

15.4 Criticism

Critics of the system have pointed out that the system described by Heins simply demonstrates a change in the motor's hysteresis drag, increasing the speed of the rotor but not producing any energy.[12] In other words, when the rotor exhibits acceleration following a specific electrical short-out, the device is merely more efficiently converting the input electricity to mechanical energy than in the other test configurations.

On February 29, 2008, six members of Ottawa Skeptics, all of whom possess qualifications in engineering and technology, met at the Colonel By building at the University of Ottawa to witness a demonstration of Perepiteia. Heins, who conducted the demonstration, later met with the members to discuss his device and answer questions. In a subsequent report released in May, Ottawa Skeptics expressed severe doubts about Heins' claims regarding Perepiteia. They noted that Perepiteia produces either observed acceleration or a slight increase in generator electrical output but that this alone does not automatically mean that "free energy" or perpetual motion is being produced or that there is a "real and measurable effect." While acknowledging that the speed-up behaviour of the generator cannot be fully explained, they stated that there is no evidence that Perepiteia "represents any challenge to currently known laws of physics."[13]

On May 21, 2009, a skeptic writing under the name Natan Weissman wrote an explanation of Perepiteia in relation to its motor, a Ryobi bench grinder. The author states that the acceleration behavior of the machine is due to the consumption of torque from the induction motor, rather than any unconventional manipulation of Electromagnetic fields or Counter-electromotive force.[14]

On June 3, 2013, posting in response to questions *Pure Energy Blog*, Heins provided an explanation of his claims, stating that: "A generator that requires a 1 Watt increase in mechanical drive shaft power to deliver 1 Watt of electrical power to a load would be 100% efficient. A generator that delivers 0.95 Watts with a 1 Watt increase in mechanical drive shaft power from no-load to on-load would be 95% efficient."[15]

15.5 References

[1] CA application 2437745

[2] CIPO - Patent application 2437745 - Administrative Status

[3] http://www.g9toengineering.com/backemf/demonstration.htm

[4] Free energy with magnetic reluctance

[5] The next great Canadian idea: Peripiteia generator by Sharda Prashad, Canadian Business magazine, July 11, 2008. (retrieved on January 3, 2009).

[6] Beam Me Up - Science & Science Fiction news: Thane Heins' Perepiteia device

[7] Inventor Doesn't Dare Say 'Perpetual Motion Machine', Physorg.com, February 7, 2008.

[8] Hamilton, Tyler (February 4, 2008). "Turning physics on its ear". *The Star* (Toronto). Retrieved May 22, 2010.

[9] The Chef Who Could Change The World of Physics » Popular Fidelity » Unusual Stuff

[10] Perepiteia Perpetual-Motion Machine May Actually Do...Something, gizmodo.com, February 7, 2008.

[11] An Internet commotion over 'perpetual motion', *Ottawa Citizen* (reprinted by Canada.com), March 1, 2008.

[12] Slashdot I Yet Another Perpetual Motion Device

[13] In This Town We Obey the Laws of Thermodynamics by Seanna Watson, Ottawa Skeptics May 4, 2008.

[14] Explanation of the Perepiteia rotating machine and the accompanying theory concerning "Back EMF" by Natan Weissman, SciScoop, May 21, 2009.

[15] Thane Heins Allows Open License of his Regenerative Acceleration (ReGenX) Technology by Sterling D. Allan, Pure Energy Systems News (Pure Energy Blog), June 9, 2013.

15.6 External links

- Perepiteia Generator Demonstration Video posted on YouTube.

- Videos by Thane C Heins Inventor posted on YouTube

- http://www.sciscoop.com/perepiteia.pdf

- Out of Their Minds - Episode 5 - Thane Heins, CBC Radio, July 26, 2011.

Chapter 16

Reactionless drive

A **reactionless drive** (also known by many other names, including as an *inertial propulsion engine*, a *reactionless thruster*, a *reactionless engine*, a *bootstrap drive* or an *inertia drive*) is a device to generate motion without a propellant, presumably in contradiction to the law of conservation of momentum.[1] The name comes from Newton's third law, which is usually expressed as, "for every action, there is an equal and opposite reaction." A large number of infeasible devices, such as the "Dean drive", are a staple of science fiction, particularly for space propulsion. To date, no reactionless drive has ever been validated under properly controlled conditions.

16.1 Historical attempts

Through the years there have been numerous claims for functional reactionless drive designs using ordinary mechanics (i.e. devices not said to be based on quantum mechanics, relativity or atomic forces or effects). Two of these represent their general classes: The "Dean drive" is perhaps the best known example of a "linear oscillating mechanism" reactionless drive; The "GIT" is perhaps the best known example of a "rotating mechanism" reactionless drive. These two also stand out as they both received much publicity from their promoters and the popular press in their day and both were eventually rejected when proven to not produce any reactionless drive forces. The rise and fall of these devices now serves as cautionary tale for those making and reviewing similar claims.[2]

16.1.1 Dean drive

Main article: Dean drive

The *Dean drive* was a mechanical device concept promoted by inventor Norman L. Dean. Dean claimed that his device was a "reactionless thruster" and that his working models could demonstrate this effect. He held several private demonstrations but never revealed the exact design of the models nor allowed independent analysis of them.[3][4] Dean's claims of reactionless thrust generation were subsequently shown to be in error and the "thrust" producing the directional motion was likely to be caused by friction between the device and the surface on which the device was resting and would not work in free space.[2][5]

16.1.2 Gyroscopic Inertial Thruster (GIT)

The Gyroscopic Inertial Thruster is a proposed reactionless drive based on the mechanical principles of a rotating mechanism. The concept involves various methods of leverage applied against the supports of a large gyroscope. The supposed operating principle of a GIT is a mass traveling around a circular trajectory at a variable speed. The high-speed part of the trajectory allegedly generates greater centrifugal force than the low, so that there is a greater thrust in one direction than the other.[6] Scottish inventor Sandy Kidd, a former RAF radar technician, investigated the possibility (without success) in the 1980s.[7] He posited that a gyroscope set at various angles could provide a lifting force, defying gravity.[8] In the 1990s, several people sent suggestions to the Space Exploration Outreach Program (SEOP) at NASA recommending that NASA study a gyroscopic inertial drive, especially the developments attributed to the American inventor Robert Cook and the Canadian inventor Roy Thornson.[6] In the 1990s and 2000s, enthusiasts attempted the building and testing of GIT machines.[9] Eric Laithwaite, the "Father of Maglev", received a US patent for his "Propulsion System", which was claimed to create a linear thrust through gyroscopic and inertial forces.[10] After years of theoretical analysis and laboratory testing of actual devices, no rotating (or any other) mechanical device has ever been found to produce unidirectional reactionless thrust in free space.[2]

16.2 Quasi-reactionless methods

Several kinds of thrust generating methods are in use, that are sometimes described as reactionless, because these methods do not work like rockets and reaction mass is not carried nor expelled from the spacecraft during their application. However, as such they are merely reaction-**mass**-less, but they all exchange momentum (react) with an outside agent instead.

- Electrodynamic tethers do not expel reaction mass like a rocket.[11] However, as electromagnetic fields can carry energy and momentum,[12] tethers do have a mechanism for momentum transfer, and hence are not reactionless. They react with the magnetic fields of for example a planet, and thus ultimately with the planet itself.

- Gravitational assist (sling-shot maneuver) is a field propulsion technique frequently used for interplanetary probes. The probe does not expel propellant but the interaction is not reactionless. Thrust is obtained by the spacecraft from the orbital energy of the planet when passing close by it; momentum is taken from the planet and is hence overall conserved.

- Solar sails (light sails) provide thrust by placing "sails" that reflect (or absorb) photons from a star, thus transferring momentum from the photons to the spacecraft.

- Magnetic sails and electric sails provide thrust by placing "sails" deflecting the flow of ionized particles of the solar wind by either magnetic or electric means, and thus transfer momentum from the particles to the spacecraft.

16.3 Modern approaches

Although the laws of classical physics regard reactionless propulsion as impossible, hypothetical methods based on principles from quantum mechanics, electrodynamics, relativity and nuclear physics have been put forward that would create similar effects without, the authors claim, violating any laws of physics. So far none of these methods has been unambiguously demonstrated to work in free space.

- The EmDrive is an electromagnetic-radiation-based device. One experimental device design consists of a closed asymmetric resonant cavity that is flooded with microwave radiation during operation. It is claimed that it produces reactionless thrust from the differential in the radiation pressure on the interior walls of the closed resonant cavity.

- The micronewton electromagnetic thruster is an electromagnetically powered device. A 2012 experiment was reported to produce unidirectional motion.[13]

- The Woodward effect is a hypothesis that predicts that an electromechanical device undergoing acceleration can generate a unidirectional force if certain assumptions regarding the nature of inertia are true. Experiments to conclusively demonstrate this effect have been ongoing since the 1990s.

- The quantum vacuum plasma thruster, or "Cannae drive", is a quantum-mechanics-based device. Its advocates claim it produces thrust by directing the charged particles produced by quantum vacuum fluctuations with electromagnetic fields. As such it wouldn't be truly reactionless, but its effect, should it be proven to work, would be similar to a reactionless device. An experimental device was tested by a NASA researcher in 2014, who claimed it produced anomalous readings inconsistent with standard physics.

16.3.1 Devices that do not generate thrust

Because there is no well-defined "center of mass" in curved spacetime, general relativity allows a stationary object to, in a sense, "change its position" in a counter-intuitive manner, without violating conservation of momentum.

- The Alcubierre drive is a hypothetical method of propulsion postulated from the theory of general relativity. Although this concept may be allowed by the currently accepted laws of physics, it remains unproven; implementation would require a negative energy density, and possibly a better understanding of quantum gravity. It is not clear how (or if) this effect could provide a useful means of accelerating an actual space vehicle and no practical designs have been proposed, but experiments are underway at NASA's Johnson Spaceflight Center to attempt the first detection of an induced spacetime curvature, which could be the first step toward proving the validity of the concept.[14][15]

- "Swimming in spacetime" is a geometrical motive principle that exploits the curved spacetime metric of the gravitational field to permit an extended body undergoing specific deformations in shape, to change position. In weak gravitational fields, like that of Earth, the change in position per deformation cycle would be far too small to detect, but the concept remains of interest as the only unambiguous example of reactionless motion in mainstream physics.[16][17]

16.4   See also

- Abraham–Minkowski controversy

- Beam-powered propulsion

- Bernard Haisch

- Field propulsion

- Harold E. Puthoff

- Inertialess drive

- Perpetual motion

- Spacecraft propulsion

- Stochastic electrodynamics

16.5   References

[1] Plait, Phil. "NASA's Quantum Drive: Cool Your Jets". *Phil Plait's Bad Astronomy blog, via Slate.* Retrieved 25 November 2014.

[2] Mills, Marc G.; Thomas, Nicholas E. (July 2006). *Responding to Mechanical Antigravity* (PDF). 42nd Joint Propulsion Conference and Exhibit. NASA. Archived from the original (PDF) on 2011-10-30.

[3] "Engine With Built-in Wings". *Popular Mechanics.* Sep 1961.

[4] "Detesters, Phasers and Dean Drives". *Analog.* June 1976.

[5] Goswami, Amit (2000). *The Physicists' View of Nature.* Springer. p. 60. ISBN 0-306-46450-0.

[6] LaViolette, Paul A. (2008). *Secrets of Antigravity Propulsion: Tesla, UFOs, and Classified Aerospace Technology.* Inner Traditions / Bear & Co. p. 384. ISBN 1-59143-078-X.

[7] *New Scientist* **148**: 96. 1995. Missing or empty |title= (help)

[8] Childress, David Hatcher (1990). *Anti-Gravity & the Unified Field.* Lost Science. Adventures Unlimited Press. p. 178. ISBN 0-932813-10-0.

[9] "The Adventures of the Gyroscopic Inertial Flight Team". 1998-08-13.

[10] U.S. Patent 5,860,317

[11] Tethers | Macmillan Space Sciences. Accessed 2008-05-04.

[12] "Special Projects Group via Internet Archive. Accessed 2008-05-04". Web.archive.org. 2002-11-13. Retrieved 2011-06-21.

[13] Charrier, Dimitri S.H. (July 18, 2012). "Micronewton electromagnetic thruster". *Applied Physics Letters* (American Institute of Physics) **101**: 034104. Bibcode:2012ApPhL.101c4104C. doi:10.1063/1.4737940. Retrieved January 4, 2014.

[14] Kakaes, Konstantin. "Warp Factor: A NASA scientist claims to be on the verge of faster-than-light travel: is he for real?, Popular Science, April 2013". PopSci.com. Retrieved 2014-11-22.

[15] http://ntrs.nasa.gov/archive/nasa/casi.ntrs.nasa.gov/ 20110015936.pdf

[16] http://www.nature.com/scientificamerican/journal/v301/ n2/full/scientificamerican0809-38.html

[17] "Swimming Through Empty Space". *Science 2.0.*

16.6   External links

- "Breakthroughs" commonly submitted to NASA

- Inertial Propulsion Engine

- Reactionless Propulsion (Not) at MathPages

Chapter 17

Charles Redheffer

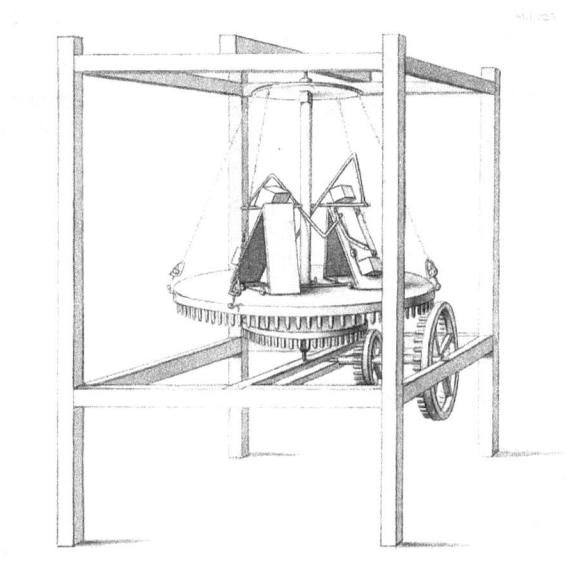

A diagram of Redheffer's first machine

Charles Redheffer was an American inventor who claimed to have invented a perpetual motion machine. First appearing in Philadelphia, Redheffer exhibited his machine to the public, charging high prices for viewing. When he applied to the government for more money, a group of inspectors were sent to examine the machine. It was discovered the machine was actually powered by a device Redheffer claimed was powered by the machine.

Redheffer moved to New York City and set up a similar scam after rebuilding his machine. However, an engineer detected that it was a fake when he visited an exhibition by listening to its unsteady motions. He discovered that the machine was operated by a man using a crank in a room on the floor above. Redheffer returned to Philadelphia. He later claimed to have created another machine, but refused to demonstrate it to anyone. He managed to get a patent for his machine in 1820, but after this his fate is unknown.

17.1 Personal life

Little has been recorded about Redheffer's life, other than his connection to the hoax. According to one source, he was from Germantown in Philadelphia,[1] but most sources simply state that he appeared in Philadelphia with his machine. Redheffer disappeared from public view after the discovery of the fraud, and his fate is unknown.

17.2 Appearance in Philadelphia

Charles Redheffer and his machine became well known in Philadelphia in 1812.[2] Redheffer claimed he had invented a perpetual motion machine and exhibited it in a house near the Schuylkill River in the outskirts of the city.[3] He charged an admission fee of $5 (some sources claim $1) for men to view it; depending on the source, women were admitted free or at a charge of $1.[4][5] The machine caused a sensation, and Redheffer lobbied for funds to build a larger version.[3]

On January 21, 1813, eight city commissioners visited Redheffer to inspect the machine. They had to do so through a barred window,[3] as Redheffer was concerned anyone going near the machine might damage it.[6] One of the inspectors, Nathan Sellers, was accompanied by his son Coleman, who noticed something odd about the gears. The machine itself was said to be powering a separate device through a series of gears and weights. Coleman noticed that the cogs were worn on the wrong side and suggested that the device was in fact powering the machine.[7]

The elder Sellers was convinced the machine was a hoax. To validate his suspicions, he hired local engineer Isaiah Lukens to build a similar machine, using a hidden clockwork motor as a power source.[8] They then arranged a demonstration of the machine to Redheffer, who was immediately convinced and offered to buy it.[7] Meanwhile, Redheffer's machine appeared in the *Philadelphia Gazette*. Civil engineer Charles Gobort offered to bet sums of money ranging from $6,000 to $10,000 that the machine was genuine, and that Redheffer had discovered perpetual motion.[7]

17.3 Move to New York City

Robert Fulton, who discovered that the machine exhibited in New York City was a fraud

His ruse revealed, Redheffer immediately departed for New York City where he was still unknown.[6] He changed his machine somewhat so that it could not be detected as easily, and he exhibited it as he had done in Philadelphia.[9]

When mechanical engineer Robert Fulton went to see the machine, he noticed that the machine was unsteady as if someone would have powered it manually and irregularly with a crank.[6] Fulton also detected that the sound was uneven, uncharacteristic of a machine's motions. He announced the machine was a fraud, and challenged Redheffer exclaiming he would expose the secret power source, otherwise he would pay for all the damage he would cause. Redheffer agreed, so Fulton removed some boards from the wall alongside the machine and exposed a catgut cord that led to the upper floor. Upstairs he found an old man who was turning a hand-crank with one hand and eating bread

with the other. Spectators realized they had been duped and destroyed the machine; Redheffer fled the city.[7]

17.4 Later appearances

Redheffer appears to have constructed another machine in 1816, which he stated his intention to demonstrate to a group of men including the mayor and chief justice of Philadelphia. However, despite several meetings, Redheffer refused to demonstrate the machine to them.[10]

On July 11, 1820, the U.S. Patent Office granted a patent to Charles Redheffer (or Charles Redheiffer) for a device listed as "machinery for the purpose of gaining power".[11] (Unfortunately, all patents up to 1836 were lost in the 1836 U.S. Patent Office fire. If recovered, it would be X-Patent X3,215.)

17.5 Notes

[1] Weiss, p.35

[2] Ord-Hume, p.125

[3] Ord-Hume, p.126

[4] Ord-Hume, p.130

[5] "The Puzzle of Perpetual Motion". *The Sydney Morning Herald* (Fairfax Media). 26 July 1954. p. 8.

[6] "Perpetual Motion Machine of Charles Redheffer". Museum of Hoaxes. Retrieved 28 August 2009.

[7] Hicks, Clifford B. (April 1961). "Why won't they work?". *American Heritage Magazine* (American Heritage Publishing Company). Retrieved 28 August 2009.

[8] Ord-Hume, p.127

[9] Ord-Hume, p.132

[10] Ogden-Niles, p.26-27

[11] Force, p.145

17.6 References

- Force, Peter (1821). *A National Calendar ... Volume 2*. Davis and Force.

- Ogden Niles, William (1817). *Niles' weekly register, Volume 11*. H. Niles.

- Ord-Hume, Arthur W. J. G. (2006). *Perpetual Motion: The History of an Obsession.* Adventures Unlimited Press. ISBN 978-1-931882-51-4.

- Weiss, Harry B.; Ziegler, Grace M. (1978). *Thomas Say, early American naturalist.* Ayer Publishing. ISBN 978-0-405-10737-5.

Chapter 18

Space-time crystal

A **space-time crystal** or **four-dimensional crystal** is a theoretical structure periodic in time and space. It extends the idea of a crystal to four dimensions.[1][2] The idea was proposed by Frank Wilczek in 2012. His speculation was that a construct would have a group of particles that move and periodically return to their original state, perhaps moving in a circle, and form a time crystal. In order for this perpetual motion to work, the system must not radiate its rotational energy.[3] This type of motion is distinct from that of persistent currents in a superconductor, wherein the rotating Cooper pairs are not time crystals because their wave functions are homogenous, meaning time translational symmetry isn't broken.[4] Symmetry would be spontaneously broken in Wilczek's ring system if its ground state still involves continuous movement.

Tongcang Li and others proposed a system with beryllium ions circulating in a magnetic ion trap at about 10^{-9} K.[4] Wilczek also suggested that a computing device could be possible with different rotational states representing information, and maybe different kinds of ions. Since this construct is in the lowest energy state it could in principle survive the heat death of the universe and continue forever.[5]

In May 2013 researchers announced they will attempt to build a component of a space time crystal, by making a rotating ring of calcium ions. Their location will be confined by electric field, and rotation in a ground state will be forced by a magnetic field. Unwanted disturbances will be minimized by reducing the temperature to 1 μK by way of laser cooling. The ion trap will be 100 μm wide. Possible rotation of the ion ring will be demonstrated by using a laser to electronically excite one of the trapped ions.[6]

Patrick Bruno has criticized this concept, arguing that Wilczek's rotating state is not the ground state of the system. He derives the supposed true, non-rotating ground state.[7] In August 2013 Bruno presented arguments that indicated rotating ground-state systems are impossible.[8]

Haruki Watanabe and Masaki Oshikawa formalised the definition of space time crystals from a ground state only requirement to also include states in thermal equilibrium. The definition used the correlation of the local order parameter at different points in space and time. This correlation in a time crystal shows a periodic oscillation as a function of time difference even as volume is increased to infinity. Next they claimed to show that time translation symmetry cannot be broken thus proving that time crystal do not exist. With the extension of the definition to crystals with a finite temperature, the Lieb-Robinson bound is used to show that for small enough time intervals the correlation over a time difference has an upper bound that tends to 0 as the volume increases.[9][10]

18.1 References

[1] Yirka, Bob (9 July 2012). "Physics team proposes a way to create an actual space-time crystal". *Phys.org*. Retrieved 15 July 2012.

[2] Wolchover, Natalie (25 April 2013). "Perpetual Motion Test Could Amend Theory of Time". The Simons Foundation. Retrieved 29 April 2013.

[3] Kentucky, FC (26 June 2012). "How to Build A Space-Time Crystal". *Technology Review*. MIT. Retrieved 18 July 2012.

[4] arXiv:1206.4772 Tongcang Li, Zhe-Xuan Gong, Zhang-Qi Yin, H. T. Quan, Xiaobo Yin, Peng Zhang, L.-M. Duan, Xiang Zhang. *Phys. Rev. Lett.* 109, 163001. 21 June 2012 "Space-time crystals of trapped ions".

[5] Aron, Jacob (6 July 2012). "Computer that could outlive the universe a step closer". *New Scientist*. Retrieved 17 July 2012.

[6] Hewitt, John (4 May 2013). "Creating time crystals with a rotating ion ring". Retrieved 4 May 2013.

[7] Bruno, Patrick (March 2013). "Comment on "Quantum Time Crystals"". *Phys. Rev. Lett.* **110** (11): 118901. Bibcode:2013PhRvL.110k8901B. doi:10.1103/PhysRevLett.110.118901. Retrieved 28 April 2013.

[8] Bruno, Patrick (August 2013). "Impossibility of Spontaneously Rotating Time Crystals: A No-Go Theorem". *Phys. Rev. Lett.* **111** (07): 070402. arXiv:1306.6275. Bibcode:2013PhRvL.111g0402B. doi:10.1103/PhysRevLett.111.070402. Retrieved 26 November 2013.

[9] Lisa Zyga (9 July 2015). "Physicists propose new definition of time crystals—then prove such things don't exist".

[10] Watanabe, Haruki; Oshikawa, Masaki (24 June 2015). "Absence of Quantum Time Crystals". *Physical Review Letters* **114** (25). doi:10.1103/PhysRevLett.114.251603.

18.2 Further reading

- H. Brown, R. Bulow, J. Neubuser, H. Wondratschek and H. Zassenhaus, *Crystallographic Groups of Four-Dimensional Space*. Wiley, New York, 1978

- Toffoli, Tommaso (2004). "A pedestrian's introduction to spacetime crystallography". *IBM Journal of Research and Development* **48** (1): 13–29. doi:10.1147/rd.481.0013. ISSN 0018-8646.

- Frank Wilczek. "Time Crystals" (PDF). Retrieved 19 July 2012.

- Frank Wilczek 11 July 2012 *Quantum Time Crystals*

- Alfred Shapere; Frank Wilczek (12 July 2012). "Classical Time Crystals". 2.

Chapter 19

Steorn

Steorn Ltd /ˈstjɔrn/ is a small, private technology development company based in Dublin, Ireland. It announced in August 2006 it had developed a technology which provides "free, clean, and constant energy," apparently in violation of the law of conservation of energy,[3] a fundamental principle of physics.[4]

Steorn challenged the scientific community to investigate their claim[5] and, in December 2006, said that it had chosen a jury of scientists to do so.[6] In June 2009 the jury gave its unanimous verdict that Steorn had not demonstrated the production of energy.[7]

Steorn has also given two public demonstrations of their technology. In the first demonstration, in July 2007 at the Kinetica Museum in London, the device failed to work.[8] The second demonstration, which ran from December 2009 to February 2010 at the Waterways Visitor Centre in Dublin, involved a motor powered by a battery and provided no independent evidence that excess energy was being generated.[9] It was dismissed by the press as an attempt to build a perpetual motion machine,[10] and a publicity stunt.[11]

Beginning in December 2015, Steorn began accepting orders for two products, through email only. The announcement was posted only to a Facebook page titled "Orbo" and a Steorn YouTube channel.[12] In early December, Steorn CEO Shaun McCarthy said that he was waiting for the first shipment of the two products, the Orbo Phone and the Orbo Cube, from a manufacturer in China.[13]

19.1 History

Steorn was founded in 2000[14] and, in October 2001, their website stated that they were a "specialist service company providing programme management and technical assessment advice for European companies engaging in e-commerce projects". Steorn is a Norse word meaning to guide or manage.

In May 2006, *The Sunday Business Post* reported that Ste-

orn was a former dot-com company which was developing a microgenerator product based on the same principle as self-winding watches, as well as creating e-commerce websites for customers. The company had also recently raised about €2.5 million from investors and was three years into a four-year development plan for its microgenerator technology.[2] Steorn has since stated that the account given in this interview was intended to prevent a leak regarding their free energy technology.[15]

The company's investment history shows several share allotments for cash between August 2000 and October 2005,[16] the investments totalling €3 million.[2] In 2006, Steorn secured €8.1 million in loans from a range of investors in order to continue their research, and these funds were also converted into shares.[17] Steorn said that they would seek no further funding while attempting to prove their free-energy claim in order to demonstrate their genuine desire for validation.[17]

19.2 Free energy claim

In August 2006, Steorn placed an advertisement in *The Economist* saying that they had developed a technology that produced "free, clean and constant energy".[5] Called Orbo, the technology was said to violate conservation of energy[3] but had been validated by eight independent scientists.[18] None of these scientists would talk to the media, and Steorn suggested that this was because they did not want to become embroiled in a controversy.[18]

19.2.1 Views on the technology

No specific details of the workings of the claimed technology have been made public. Seán McCarthy stated in a 2006 RTÉ radio interview, "What we have developed is a way to construct magnetic fields so that when you travel round the magnetic fields, starting and stopping at the same position, you have gained energy".[19] In 2011, Steorn's website was updated to suggest that the Orbo is based on magnetic fields which vary over time.[20] Barry Williams of the Australian Skeptics has pointed out that Steorn is "not the first company to claim they have suddenly discovered the miraculous property of magnetism that allows you to get free energy"[4] while Martin Fleischmann says that it is not credible that positioning of magnetic fields could create energy.[18]

Following a meeting between McCarthy and Professor Sir Eric Ash in July 2007, Ash reported that "the *Orbo* is a mechanical device which uses powerful magnets on the rim of a rotor and further magnets on an outer shell".[21] During this meeting, McCarthy referred to the law of conservation of energy as scientific dogma.[21] However, conservation of energy is a fundamental principle of physics,[4] more specifically a consequence of the unchanging nature of physical laws with time by Noether's Theorem. Ash said that there was no comparison with religious dogma since there is no flexibility in choosing to accept that energy is always conserved.[21] Rejecting conservation of energy would undermine all science and technology.[21] Ash also formed the opinion that McCarthy was truly convinced in the validity of his invention but that this conviction was a case of "prolonged self-deception".[21]

Many people have accused Steorn of engaging in a publicity stunt although Steorn denies such accusations.[22] Eric Berger, writing on the *Houston Chronicle* website, commented: "Steorn is a former e-business company that saw its market vanish during the dot.com bust. It stands to reason that Steorn has retooled as a Web marketing company and is using the "free energy" promotion as a platform to show future clients how it can leverage print advertising and a slick Web site to promote their products and ideas".[11] Thomas Ricker at *Engadget* suggested that Steorn's free-energy claim was a ruse to improve brand recognition and to help them sell Hall probes[23] while Josh Catone, features editor for *Mashable*, believes that it was merely an elaborate hoax.[24]

19.2.2 Jury process

In its advertisement in *The Economist*, Steorn challenged scientists to form an independent jury to test their technology and publish the results.[25][26] Within 36 hours of the advertisement being published, 420 scientists contacted Steorn[27] and, on 1 December 2006, Steorn announced it had selected a jury.[6] It was headed by Ian MacDonald, emeritus professor of electrical engineering at the University of Alberta, and the process began in February 2007.[7]

In June 2009 the jury announced its unanimous verdict that "Steorn's attempts to demonstrate the claim have not shown

the production of energy. The jury is therefore ceasing work".[7] Dick Ahlstrom, writing in the *Irish Times*, concluded from this that Steorn's technology did not work.[7] Steorn responded by saying that because of difficulties in implementing the technology the focus of the process had been on providing the jury with test data on magnetic effects for study.[28] Steorn also said that these difficulties had been resolved and disputed its jury's findings.[7][28]

19.2.3 Demonstrations

A notice at the Kinetica Museum announcing the cancellation of the public demonstration

On 4 July 2007, the technology was to be displayed at the Kinetica Museum, Spitalfields Market, London. A unit constructed of clear plastic was prepared so that the arrangement of magnets could be seen and to demonstrate that the device operated without external power sources.[8][29] The public demonstration was delayed and then cancelled because of technical difficulties. Steorn initially said that the problems had been caused by excessive heat from the lighting[8][30] but later blamed the failure on damage done to bearings due to a greenhouse effect within the box.[31]

A second demonstration ran between 15 December 2009 and February 2010[32] at the Waterways Visitor Centre in Dublin, and was streamed via Steorn's website.[33][34] The demonstration was of a device powered by a rechargeable battery. Steorn said that the device produced more energy than it consumed and recharged the battery.[9] No substantive details of the technology were revealed and no independent evidence of Steorn's claim was provided.[9]

On 1 April 2010 Steorn opened an online development community, called the Steorn Knowledge Development Base (SKDB), which they said would explain their technology.[35] Access is available only under licence on payment of a fee.[35][36]

In May 2015, Steorn put an "Orbo PowerCube" on display behind the bar of a pub in Dublin. The PowerCube was a small box which the pub website claimed contained a "perpetual motion motor" which required no external power source. The cube was shown charging a mobile phone. Steorn claimed to be performing some "basic field trials" in undisclosed locations.[37]

19.2.4 Products

On 29 September 2015, Steorn announced that it would soon begin a series of live online webinars showcasing the Orbo technology. The first in the series of webinars took place on 28 October 2015, and the second on 1 December 2015.[38] The first webinar revealed a product titled *OCube* which is a small device with a USB port charging smartphones and computer tablets. The second webinar revealed how to order the O-Cube and also a new mobile phone, the *OPhone*, which Steorn purports as being able to operate without ever needing to be externally charged.[12]

19.3   See also

- History of perpetual motion machines

19.4   References

[1] "Steorn Investor Relations". Steorn Ltd. 9 February 2006. Retrieved 11 September 2007.

[2] Daly, Gavin (21 May 2006). "Firm strives to extend mobile battery lifespans". *ThePost.IE*. Retrieved 25 October 2006.

[3] "Our Claim". Steorn Ltd. Archived from the original on 2 May 2007. Retrieved 12 April 2007.

[4] Weekes, Peter (20 August 2006). "Irish energy miracle 'a joke'". Melbourne: The Age. Retrieved 20 August 2006.

[5] "Copy of Steorn advertisement featured in The Economist, hosted by dispatchesfromthefuture.com" (JPEG). Retrieved 21 January 2009.

[6] "Steorn finalises contracts for jury to test its free energy technology". Steorn (archive copy from archive.org). 1 December 2006. Archived from the original on 2007-02-21. Retrieved 5 March 2009.

[7] Dick Ahlstrom (24 June 2009). "Irish "energy for nothing" gizmo fails jury vetting". *Irish Times*. (subscription required)

[8] "Irish firm's display of 'free-energy' machine delayed". *Belfast Telegraph*. 5 July 2007.

[9] Rupert Goodwins (15 December 2009). "Steorn shows revolving Orbo to the public". ZDNet. Retrieved 15 December 2009.

[10] Goldacre, Ben (7 July 2007). "Perpetual motion goes into reverse". *The Guardian*. Retrieved 10 June 2013.

[11] Berger, Eric (19 August 2006). "Steorn and free energy: the plot thickens". *SciGuy. Houston Chronicle* blogs. Retrieved 21 August 2006.

[12] McCarthy, Shaun (1 December 2015). "Orbo Webinar 2". *Steorn*. Retrieved 1 December 2015.

[13] McNulty, Paul. "In the Docklands, a Company Relaunches Claims of Perpetual Motion Machine". *Dublin Inquirer*. Retrieved 2015-12-09.

[14] "Wanted: scientists to test free energy technology". *Irish Examiner*. 20 August 2006. Archived from the original on 2006-08-21. Retrieved 20 August 2006.

[15] "Energy Issues". Steorn. 1 October 2006. Retrieved 26 October 2006.

[16] "Steorn Company Submissions". Companies Registration Office. Retrieved 16 October 2006.

[17] Downes, John (10 August 2008). "'Free energy' firm generated €8m in funding". *Sunday Tribune*. Retrieved 5 November 2008.

[18] Boggan, Steve (25 August 2006). "These men think they're about to change the world". *The Guardian* (London). Retrieved 24 May 2010.

[19] "Irish company challenges scientists to test 'free energy' technology". *Yahoo! News*. 18 August 2006. Archived from the original on 3 September 2006.

[20] "Orbo". Steorn Ltd. Archived from the original on 16 July 2011. Retrieved 18 November 2011.

[21] "The perpetual myth of free energy". *BBC News*. 9 July 2007. Retrieved 9 July 2007.

[22] Chris Vallance (23 August 2006). "Caught in a Tale Spin". *Pods&Blogs*. BBC. Retrieved 25 June 2009.

[23] Thomas Ricker (25 June 2009). "Steorn gives up on free-energy, starts charging for USB-powered divining rods". Engadget. Retrieved 25 June 2009.

[24] Catone, Josh (15 July 2009). "Top 15 Web Hoaxes of All Time". Mashable. Retrieved 21 July 2009.

[25] "Steorn develops free energy technology and issues challenge to the global scientific community". Steorn Ltd. 18 August 2006. Retrieved 29 June 2009.

[26] "Steorn announces plans for widespread deployment of its free energy technology post-validation". Steorn. 11 January 2007. Retrieved 6 July 2007.

[27] Smith, David (20 August 2006). "Scientists flock to test 'free energy' discovery". London: Guardian Unlimited. Retrieved 20 August 2006.

[28] "Jury report". June 2009. Archived from the original on 2010-12-30.

[29] "'Free' energy technology goes on display". *The Irish Times*. 4 July 2007. Retrieved 5 July 2007.

[30] "Steorn announcement: Kinetica Demonstration". 6 July 2007. Retrieved 5 June 2007.

[31] Schirber, Michael (August 2007). "Harsh light shines on free energy". *Physics World* **20** (8): 9.

[32] "Testing - Orbo Technology Update". Steorn. 11 February 2010. Retrieved 13 February 2010.

[33] Rupert Goodwins (14 December 2009). "Steorn sets up for second bite at perpetual cherry". ZDNet. Retrieved 14 December 2009.

[34] "Steorn Announces Public Demonstration of Orbo Technology". Steorn. 15 December 2009. Retrieved 15 December 2009.

[35] "SKDB Launch". Steorn. 1 April 2010. Retrieved 9 May 2010.

[36] Gavin Daly (6 June 2010). "'Free' energy firm to make over €2m this year". ThePost.ie. Retrieved 8 June 2010.

[37] Boran, Marie (14 May 2015). "Self-charging battery causes a stir in Dublin pub test". Irish Times. Retrieved 17 June 2015.

[38] Brown, Craig (29 Sep 2015). "EXCLUSIVE: LIVE Steorn Webinars on Orbo PowerCube". Free Energy News. Retrieved 13 Oct 2015.

19.5 External links

- Steorn official website (now a Facebook page)

- "7 minute television interview with Seán McCarthy". *Sky News*. 8 September 2006.

- Search for Steorn patent applications

- Michael Hanlon (28 August 2006). "Have scientists unlocked the secret of perpetual motion?". *Daily Mail* (London). Retrieved 8 March 2009.

- "Free, Clean Energy For All?". *ABC News*. 23 August 2006.

- Freeman, John (May 2008). "The Steorn Exploit and its Spin Doktors, or "Synergie ist der name of das Spiel, my boy!"". *Postmodern Culture* **18** (3). Direct link to article

19.6 Text and image sources, contributors, and licenses

19.6.1 Text

- **Perpetual motion** *Source:* https://en.wikipedia.org/wiki/Perpetual_motion?oldid=694227904 *Contributors:* Derek Ross, Bryan Derksen, Robert Merkel, The Anome, Tarquin, Miguel~enwiki, Roadrunner, SimonP, Maury Markowitz, David spector, Heron, ChrisSteinbach, Someone else, Xoder, D, DrewT2, Michael Hardy, Tim Starling, Ezra Wax, Liftarn, MartinHarper, Gabbe, Bobby D. Bryant, Karada, Minesweeper, Card~enwiki, Ihcoyc, Ellywa, DavidWBrooks, Haakon, William M. Connolley, Palfrey, Evercat, Vodex, Charles Matthews, Dcoetzee, Ed Cormany, Reddi, Stismail, Furrykef, VeryVerily, Omegatron, Goose, Bcorr, Pakaran, Finlay McWalter, EpiVictor, Robbot, Blainster, Paul Murray, Cyrius, Jooler, Carnildo, Pablo-flores, Wjbeaty, Giftlite, Dbenbenn, Smjg, Jao, Jyril, Wolfkeeper, Tom harrison, Ferkelparade, Timpo, Karn, Abqwildcat, Leonard G., Jfdwolff, Duncharris, Tagishsimon, Edcolins, Saxsux, Knutux, Abu badali, Antandrus, Beland, WhiteDragon, Rdsmith4, Burgundavia, Elroch, Icairns, Nielmo~enwiki, Indolering, Agarner, JHCC, Joyous!, TJSwoboda, Cky, Zro, Mike Rosoft, Miborovsky, Poccil, Eb.hoop, Rich Farmbrough, Pjacobi, Vsmith, Izwalito~enwiki, Dave souza, Xezbeth, CABAL, Bender235, Kjoonlee, Plugwash, Violetriga, Pt, El C, RoyBoy, Femto, Jon the Geek, JRM, Kghose, Stesmo, Enric Naval, Viriditas, .:Ajvol:., Foobaz, Reuben, Neg, Acjelen, Physicistjedi, Ultra megatron, Gunter.krebs, Alansohn, Cammoore, Rgclegg, Ashley Pomeroy, AzaToth, Apoc2400, CJ, Tom12519, Hohum, MoraSique, Mahtan, Wtmitchell, Evil Monkey, Gortu, MIT Trekkie, Johntex, Kazvorpal, Adrian.benko, Bobrayner, Blaze Labs Research, Richard Arthur Norton (1958-), Simetrical, Firsfron, RHaworth, Miaow Miaow, Hdante, Wikiklrsc, GregorB, Isnow, Hayvac, Hughcharlesparker, Prashanthns, Rusty2005, Cshirky, Gerbrant, Seb-Gibbs, Sjakkalle, Rjwilmsi, Nanami Kamimura, Erebus555, Jake Wartenberg, Mixel~enwiki, Thomas Arelatensis, Gareth McCaughan, NeonMerlin, Moorlock, LjL, Bubba73, Dianelos, Fred Bradstadt, Ian Pitchford, Arnero, Margosbot~enwiki, Skierpage, SteveBaker, Srleffler, Kelpi, Antimatter15, Volunteer Marek, Zhipengsun, YurikBot, Wavelength, Splintercellguy, Hairy Dude, Kencaesi, Jachin, Waitak, Hillman, Bobby1011, Jurijbavdaz, DMahalko, Robert A West, Bhny, Lar, Alex Ramon, Cryptic, Bovineone, Salsb, Nowa, Test-tools~enwiki, Janus20, Długosz, Belfroy, Pyroclastic, Scs, Jiks, Jeh, Silverhill, Mattcolville, Amnewsboy, Salt3d, AvalonXQ, Light current, Enormousdude, 21655, 2over0, Closedmouth, Tevildo, JoanneB, Chrishmt0423, Smurrayinchester, Frnknstn, Infowarrior, Monk of the highest order, Somejeff, Rpupkin77, Aerno, Sbyrnes321, DVD R W, CIreland, SmackBot, Prebys, Tumbleman, Bswee, Iacobus, Rex the first, Paulkramer, Anarchist42, InverseHypercube, KnowledgeOfSelf, McGeddon, Ze miguel, CyclePat, Lawrencekhoo, Jacek Kendysz, Jagged 85, Ssbohio, WookieInHeat, RedSpruce, Alksub, CrypticBacon, Gilliam, Betacommand, BBMSteve, Kmarinas86, Chris the speller, Wuffyz, Isaacsurh, MalafayaBot, Hieros, Sbharris, Zven, MetS-Energie, Can't sleep, clown will eat me, Nick Levine, Shalom Yechiel, Fraser Chapman, OrphanBot, Nixeagle, Nima Baghaei, LeContexte, Addshore, Threeafterthree, Xichael, Aldaron, Wen D House, Cybercobra, Bowlhover, InAJar, Pwjb, Bronzie, BullRangifer, Mwtoews, Jklin, DMacks, Henning Makholm, Jmak, Itmozart, Byelf2007, Saccerzd, Rklawton, SatsukiMikata, Vanished user 9i39j3, Tenebrous, Minglex, Stelio, Flamboyant~enwiki, UKER, Panaceus, Zapvet, Coaxial, ShakingSpirit, BranStark, Iridescent, Michaelbusch, Newone, MOBle, Not my leg, GDallimore, Casper Gutman, Audiosmurf, Pinbucket, Nuttyskin, Tawkerbot2, Chetvorno, Mellery, CmdrObot, TheHerbalGerbil, Van helsing, Picaroon, Cartoonmaster, Banedon, Jaxad0127, SEJohnston, Bakanov, Smoove Z, Esahr, The Isiah, ProfessorPaul, Revolus, Cydebot, Reywas92, Louiegarcia 3@hotmail.com, Cyhawk, Gogo Dodo, JFreeman, Uker, A Softer Answer, Lugnuts, Wfaxon, HitroMilanese, Starionwolf, NMChico24, Reyalicea, Smiteri, Epbr123, Jthinker23, Cetinsert, Pajz, Gromonger-17, Inanimous, Headbomb, Second Quantization, Captain Crawdad, MichaelMaggs, Perpetual motion machine, Pgagge, Mentifisto, WikiSlasher, AntiVandalBot, Fireplace, MoogleDan, Courtjester555, Bobbfwed, LuckyLouie, Justinmeister, Random user 8384993, Zedla, Hexc0de, Toastydeath, Canadian-Bacon, Phil153, Narssarssuaq, MER-C, Csaboka, SiobhanHansa, Jcmac, Magioladitis, Bongwarrior, VoABot II, Vintei, James-BWatson, Rivertorch, Jim Douglas, Rich257, Homunq, Catgut, Neilljones, Gunsfornuns, Felliax08, Silentaria, Glen, DerHexer, WLU, Robin S, Will2green, DGG, Gwern, Stephenchou0722, Ssivaa, PsyMar, IgorSF, Eferen, Dkrogers, Christian.Mercat, Discpad, Huzzlet the bot, Numbo3, Maurice Carbonaro, Gasmaster2000, Yishaika, Amckeen, PerpetuumMobile, Patentinfo, McSly, Azndragon1987, Kerem Ozgur, Cometstyles, DorganBot, Alwynjunior, Squids and Chips, Jeff G., Indubitably, Antiarchangel, Af648, TXiKiBoT, Jkstark, IPSOS, OlavN, MustardDog, Jhkey, Steven J. Anderson, Voiceofreason01, Foltinek12, Tsob, Raymondwinn, BotKung, Dick Kimball, GaylordBumBum, Sapphic, Dustybunny, Mike4ty4, Bernstein2291, Legoktm, SaltyBoatr, SieBot, Tomalak geretkal, PlanetStar, Pi is 3.14159, Jpelton, Digwuren, Toddst1, Jojalozzo, Oxymoron83, Reginmund, Blacklemon67, Thatotherdude, WikiLaurent, Stabguy, Jonathanstray, Jehkque, ClueBot, Amanmore.96, Hartiberlin, 0nullbinary0, Dudemonkeys, Othmanskn, Mekkel.Richards, Mr. Laser Beam, CohesionBot, Alexbot, Rhododendrites, Sun Creator, Juanmlleras, Cula, Wdford, Coccyx Bloccyx, Aitias, Introductory adverb clause, NJGW, Johnuniq, SoxBot III, Alastair Carnegie, DumZiBoT, BarretB, XLinkBot, Hotcrocodile, Vayalir, Flumstead, Abbas13677 2005, Rror, Little Mountain 5, Kajabla, Addbot, C6541, Some jerk on the Internet, Donhoraldo, Glane23, Hanch-Bibi, Djcustomcomputing, Ehrenkater, Verbal, Go-here.nl, Lightbot, Guyonthesubway, TundraGreen, Legobot, Luckas-bot, Yobot, Gdewilde, Armchair info guy, AnomieBOT, Jim1138, Piano non troppo, Typer in Time, Materialscientist, Citation bot, Xqbot, Bihco, Ched, GrouchoBot, Ozirock, Chjoaygame, FrescoBot, Tobby72, George Mel, EFFemeer, Sa3er2, Jonathansuh, Sylvester107, Kwiki, David77graham, Citation bot 1, HRoestBot, EM00 wiki, ARCWIKI, Josus chroates, GregoryGOrDon, Sonofaqua, MrX, Maxspawn5, Dalba, Jackehammond, DoRD, EmausBot, Sir Arthur Williams, Themastertree, Rami radwan, Openlysolven, K6ka, Slawekb, Solomonfromfinland, MCMLXXXVII, Knight1993, Patsobest, ErratumMan, Sbmeirow, Citron, Brandmeister, Lawstubes, DennisIsMe, Teapeat, Amruth M D, Rememberway, ClueBot NG, Joefromrandb, Rezabot, Helpful Pixie Bot, Art and Muscle, Strike Eagle, Bibcode Bot, Mdmilagre, Benzband, The Almightey Drill, TheGoodBadWorst, Ashutoshpandey123, DoctorKubla, Electricmuffin11, Khazar2, Adwaele, Dexbot, Epicgenius, FrankRadioSpecial, Comp.arch, DavRosen, Quenhitran, Oren999, UY Scuti, Bacon Lover343, Benmurray1999, The happy chicken, Kcida10, Monkbot, Apolymathman, Willmolven, Unknown125.774.979, Unknown:03487829, Asthaz, Goli.sairahul, Isambard Kingdom, Brennybob, Jerodlycett, KasparBot, Shareeffahmy, Redhair.Neve, Rodgers.Shaundine, Nathan Ngo, Datasolarenergy and Anonymous: 531

- **History of perpetual motion machines** *Source:* https://en.wikipedia.org/wiki/History_of_perpetual_motion_machines?oldid=683196842 *Contributors:* The Anome, BlckKnght, SimonP, Mjb, Heron, Camembert, Frecklefoot, Infrogmation, Sannse, Skysmith, Paul A, Alfio, DavidW-Brooks, William M. Connolley, Kvintadena, Reddi, Dysprosia, Jitse Niesen, Desertphile, Audin, Greenrd, Maximus Rex, Furrykef, Topbanana, Pstudier, Pakaran, Proteus, Finlay McWalter, Shantavira, Moriori, Fredrik, RedWolf, Flauto Dolce, Auric, Aleron235, Benc, ElBenevolente, Carnildo, Davidcannon, Alan Liefting, David Gerard, Matt Gies, Djinn112, Tom harrison, Martijn faassen, Orangemike, Zigger, Rpyle731, Duncharris, DO'Neil, Steven jones, Bobblewik, Jrdioko, Edcolins, Delta G, Golbez, Wmahan, Chowbok, R. fiend, Mihoshi, Imlepid, Icairns, Goh wz, Sam Hocevar, Joyous!, Vitaleyes, Kevyn, Kevin Rector, CALR, Discospinster, Rich Farmbrough, Guanabot, Pjacobi, Xezbeth, Mattingly23, CanisRufus, El C, Sietse Snel, Stesmo, Flxmghvgvk, Enric Naval, Cmdrjameson, I9Q79oL78KiL0QTFHgyc, Kundor, Flammifer, Polylerus, Pearle, Calton, Stillnotelf, Evil Monkey, Anthony Ivanoff, Alai, LukeSurl, Ceyockey, Sylvain Mielot, Batintherain, Kelly Martin,

- **Stefan Marinov** *Source:* https://en.wikipedia.org/wiki/Stefan_Marinov?oldid=694082995 *Contributors:* Blainster, Alan Liefting, MingMecca, Danko Georgiev, Pjacobi, Bender235, Circeus, Enric Naval, Jeltz, BillC, Rjwilmsi, Bigboehmboy, CJLL Wright, TodorBozhinov, GeeJo, Nlu, Meegs, SmackBot, Melchoir, Adrian232, LeContexte, Michael David, JzG, BillFlis, Cydebot, Ntsimp, Headbomb, Gnixon, Julia Rossi, Toohool, Dr. Submillimeter, Homo Cosmosicus, CommonsDelinker, SureFire, Jreferee, Aboutmovies, VolkovBot, Fd1969, RobertFritzius, Lightmouse, ImageRemovalBot, Niceguyedc, GregVolk, CohesionBot, Paladin R.T., Good Olfactory, Addbot, DOI bot, SpellingBot, Lightbot, Yobot, Citation bot, 78beckercolt, FrescoBot, D'ohBot, RjwilmsiBot, Solomonfromfinland, Urgent01, Bibcode Bot, VIAFbot, FrankRadioSpecial, KasparBot and Anonymous: 19

- **Maxwell's demon** *Source:* https://en.wikipedia.org/wiki/Maxwell'{ }s_demon?oldid=686903669 *Contributors:* Derek Ross, CYD, Bryan Derksen, Taral, Andre Engels, XJaM, Roadrunner, Mjb, Olivier, Dante Alighieri, 168..., Ahoerstemeier, J-Wiki, Victor Gijsbers, Michael Shields, Tristanb, Reddi, Dandrake, WhisperToMe, BenRG, Owen, Robbot, Fredrik, RedWolf, Kadin2048, Kesuari, Auric, Emyth, Johnstone, Xanzzibar, Mattflaschen, Cutler, Enochlau, Centrx, Gwalla, Aendrew~enwiki, Anville, Joconnor, Tromer, Chinasaur, DO'Neil, Pcarbonn, HorsePunchKid, Tothebarricades.tk, Icairns, Klemen Kocjancic, Mike Rosoft, Rhobite, Pjacobi, Jyp, Pavel Vozenilek, Pt, Laurascudder, WhiteTimberwolf, Pablo X, Rpresser, Dreish, Thanos6, Scott Ritchie, Wayfarer, Mennato, CyberSkull, Dudenas, Wprphd, Yamla, Ferrierd, Apoc2400, PAR, Hgrenbor, MoraSique, Jheald, Count Iblis, Vuo, Dave.Dunford, Stemonitis, Karnesky, Madmardigan53, Gruu, Mpatel, Waldir, Pfalstad, Deafgeek, Grammarbot, Rjwilmsi, Dr.Gonzo, Koavf, Barklund, Kirillz, Ligulem, Maxim Razin, Yamamoto Ichiro, Fresheneesz, NoahB, Bjrice, CiaPan, Celebere, Martin Hinks, YurikBot, Peregrine Fisher, Hairy Dude, Cyferx, Sillybilly, Hydrargyrum, Gaius Cornelius, VT~enwiki, Mipadi, Dhollm, Froth, Saberwyn, E2mb0t~enwiki, Mtcedar, Kortoso, Wknight94, CWenger, The Blue Moose, LeonardoRob0t, HereToHelp, Curpsbot-unicodify, John Broughton, ChristopherGautier, Crystallina, A bit iffy, Tttrung, SmackBot, FocalPoint, Tom Lougheed, InverseHypercube, JohnSankey, McGeddon, Antrophica, Cesoid, Sirana, ChristopherEdwards, SamWhited, Khobler, Complexica, Whispering, Sbharris, Jdthood, Shalom Yechiel, Rrburke, Richard001, Derek R Bullamore, Jbergquist, Kendrick7, N0xin, Sadi Carnot, Derekwriter, Byelf2007, Chymicus, Eliyak, Samantha of Cardyke, Erwin, Larry660, FelisSchrödingeris, Chetvorno, Mellery, Z4ns4tsu, CmdrObot, Van helsing, Shorespirit, Rawling, KyraVixen, Rogerborg, Floridi~enwiki, Nhorsley, Lgh, Myasuda, FilipeS, Cydebot, Polyesther~enwiki, UberMan5000, Clovis Sangrail, Duches77, Sthomson06, EdMercer, Thijs!bot, TheHutt, Jobber, Headbomb, Davidhorman, MichaelMaggs, MoogleDan, Danger, Ingolfson, Sluzzelin, Leuko, Barek, Bekant, Suchandsuch, B1A, Pedro, Albmont, Nessman, Jockm, Cyktsui, Ycartreel, WLU, Mrathel, CommonsDelinker, AgarwalSumeet, HEL, Huzzlet the bot, Maurice Carbonaro, RedPoptarts, PC78, Vampyrian, SlowJog, Acalamari, Collegebookworm, Notreallydavid, CompuChip, Kenneth M Burke, DorganBot, VolkovBot, Kweston, Myles325a, Jdcrutch, Enviroboy, Rubberchickenben, Paradoctor, Maxwellsdaemonuk, JohnManuel, Vandread68, BonHed, Drojem, OKBot, Tesi1700, Henry Merrivale, PaulLowrance, Veltzerdoron, Citizen Sykes, Primex~enwiki, Idleloop~enwiki, Philarete~enwiki, SchreiberBike, Lambtron, Bgeelhoed, Larrylagarto, XLinkBot, Intuit22, Trefynwy, Addbot, DOI bot, AkhtaBot, Ingeniosus, Herr Gruber, AgadaUrbanit, HandThatFeeds, Lightbot, LiteralKa, Luckas-bot, Yobot, Aldebaran66, Asieo, Sirsparksalot, AnomieBOT, Floquenbeam, Citation bot, ArthurBot, Xqbot, Ywaz, Cooltxbreeze, Scchan, Prezbo, Confront, FrescoBot, Paine Ellsworth, Plummerr, Sławomir Biały, Steve Quinn, Citation bot 1, ErStelz, Jhgheart, Full-date unlinking bot, SkyMachine, Abhaybhaskar, LilyKitty, Noieraieri, Nistra, EmausBot, KurtLC, 478jjjz, Bt8257, Solomonfromfinland, Josve05a, Ethaniel, Smirnoff171, Hotzemoerkerk, Unitrin, Bulwersator, Abp35, Livelydon, ClueBot NG, Rizzerwiki, Helpful Pixie Bot, Christianmetzgeer, Bibcode Bot, Smeat75, AngusWOOF, CitationCleanerBot, Snow Blizzard, AlanS333, Crstrode, Aleixciudad, Dexbot, Webclient101, Jasontaylor7, Kernsters, Wchristensen, Eden Fletcher, Merleblue, Crimso71, Mre env, Aho-hopc-ull, Jan Kaninchen, Tedsanders, Krotera, Monkbot, Jimph, Lessurd, Epipotamus, Araignee-pourpre and Anonymous: 254

- **Methernitha** *Source:* https://en.wikipedia.org/wiki/Methernitha?oldid=681226517 *Contributors:* MingMecca, BillC, FlaBot, Hydrargyrum, SmackBot, Hmains, Bwpach, Alaibot, Johnpacklambert, McM.bot, Vanished user ewfisn2348tui2f8n2fio2utjfeoi210r39jf, Federowicz1961, CohesionBot, Europe22, XLinkBot, Borock, Addbot, SwisterTwister, DrilBot, Weissetaube771, Lagbadi, Iranmanavi, Ocean2010, Will Beback Auto, Calabe1992, Epicgenius, Johnnybegood209, Lotuspeace and Anonymous: 6

- **Perepiteia** *Source:* https://en.wikipedia.org/wiki/Perepiteia?oldid=655786335 *Contributors:* Michael Hardy, Matt Gies, RoyBoy, Enric Naval, Firsfron, 2over0, SmackBot, Mangoe, Herostratus, LeContexte, Robofish, Cajolingwilhelm, Xionbox, GDallimore, CmdrObot, Cydebot, DumbBOT, Qwerty Binary, Sciencebase, Here2njoy, Atheuz, OlavN, Corvus cornix, Staka, Gillyweed, Bentogoa, Mr. Stradivarius, Hyperionsteel, ImageRemovalBot, DumZiBoT, XLinkBot, Wageless, Nwiggs, Reddyenumber4, TheodoreTest, Aimulti, Keepcalmandcarryon, AnomieBOT, HJ Mitchell, StephenWatson, Tbhotch, RjwilmsiBot, Solomonfromfinland, L Kensington, Bped1985, Cntras, Hardy Heck, ThaneHeins, BG19bot, ChrisGualtieri and Anonymous: 33

- **Reactionless drive** *Source:* https://en.wikipedia.org/wiki/Reactionless_drive?oldid=695192635 *Contributors:* Damian Yerrick, Skysmith, William M. Connolley, AnonMoos, Wolfkeeper, Anville, Mcapdevila, Brianhe, Xezbeth, Illumynite, Enric Naval, Elwood00, Hdeasy, BRW, Bobrayner, RHaworth, David Haslam, Mathrick, Fresheneesz, Gwernol, Hillman, Whale~enwiki, Długosz, Lexicon, Black Falcon, Zzuuzz, Petri Krohn, SmackBot, Thumperward, Pwjb, Lcarscad, Metamagician3000, ArglebargleIV, JzG, JorisvS, JHunterJ, Fangfufu, Iridescent, CmdrObot, Keraunos, Andyjsmith, Second Quantization, Davidhorman, Dbrodbeck, Malvineous, Lfstevens, Jim.henderson, Bermy88, Trilobitealive, Robin Z, Goyston, Vmaldia, Philip Trueman, Spiral5800, Andy Dingley, Jeffrey Beer, Judgeking, Sapphic, Jack Merridew, Bentogoa, Rocksanddirt, Binksternet, GorillaWarfare, DFRussia, XLinkBot, Brettpaul2000, Grav01, Addbot, Nurotoxin, OlEnglish, ATOE, Drpickem, Flash.starwalker, Yobot, RuPassenger~enwiki, Aldebaran66, Freikorp, AnomieBOT, Trevor Loughlin, Ellbug89, Cougarrcsnva, Alex Belov, Peter Bogra, Tom.Reding, Coekon, Stefan.K., Hjvanranden, Tøpholm, Ccpmny, CProvat, Russell Anderson, Uv7xhrt, Uploadvirus, Solomonfromfinland, Fæ, Ftlqed, AndyTheGrump, Rememberway, ClueBot NG, Wouldgyro, CowlishawDavid, Helpful Pixie Bot, Bibcode Bot, RovingPersonalityConstruct, BattyBot, Ggutsche1, XXN, Scie8, Rolf h nelson, Psalterion2008, Monkbot, Informedskeptic, Chrismofer and Anonymous: 81

- **Charles Redheffer** *Source:* https://en.wikipedia.org/wiki/Charles_Redheffer?oldid=618136005 *Contributors:* Skysmith, Tpbradbury, Cantthinkofagoodname, SmackBot, MartinPoulter, Cydebot, A876, Majorly, Instinct, ***Ria777, Avicennasis, Anonymous Dissident, Dwiakigle, Dabomb87, Daniel Musto, Good Olfactory, Addbot, Aiken drum, ZéroBot, Helpful Pixie Bot, MrBill3, Monkbot and Anonymous: 4

- **Space-time crystal** *Source:* https://en.wikipedia.org/wiki/Space-time_crystal?oldid=674612002 *Contributors:* BenRG, Tobias Bergemann, Graeme Bartlett, GregorB, Emerson7, Drbogdan, Wavelength, Kotra, Island Dave, Missvain, Rickard Vogelberg, Wing gundam, Flyocean, Solomonfromfinland, Black3agl353, Bibcode Bot, Mceresko, Jhareesh, Monkbot and Anonymous: 5

- **Steorn** *Source:* https://en.wikipedia.org/wiki/Steorn?oldid=694792156 *Contributors:* The Anome, Frecklefoot, Mahjongg, Kwekubo, HiramvdG, Conti, Desertphile, Greenrd, E23~enwiki, Pstudier, Jeffq, Bearcat, Robbot, Moriori, Outolumo, Sterlingda, Ludraman, Orangemike,

Karn, Zoney, Edcolins, Imlepid, Ary29, Salimfadhley, Deglr6328, Zondor, Rich Farmbrough, Pmsyyz, Pjacobi, Rannpháirtí anaithnid (old), Kjoonlee, El C, Kwamikagami, RoyBoy, Alderbourne, Smalljim, Enric Naval, Overtone, Sasquatch, Sludge, Tadman, Hdeasy, Mrholybrain, Dalillama, AiusEpsi, BigZaphod, Firsfron, Linas, LizardWizard, Professor Ninja, David Haslam, Orangehues, Nleseul, Ronnotel, Rjwilmsi, Koavf, Diza, Srleffler, WikiWikiPhil, The Rambling Man, Bhny, Limulus, Ansell, ALoopingIcon, Friday, DarthVader, Thiseye, Johantheghost, Rwxrwxrwx, 2over0, Smartaalec, Davidwt, Eric TF Bat, JDspeeder1, Somejeff, Boldra, Kgf0, Bwiki, Anthony717, SmackBot, Illuminattile, ElectricRay, Anarchist42, Hydrogen Iodide, McGeddon, Bigbluefish, Chairman S., Beest, CuriousOliver, Brianski, Autarch, Oli Filth, Timneu22, Danj205, Colonies Chris, Emurphy42, Seifip, Sparkzilla, Niall123, Ianmacm, Kirils, Mwtoews, Bezapt, JzG, Kuru, Ocee, Jaganath, Clacker, JoshuaZ, Ckatz, Baxter001, Trounce, KittensOnToast, Meco, Mets501, Spiel496, Dr.K., BananaFiend, Iridescent, Drlegendre, GDallimore, Gnome (Bot), Trade2tradewell, Will314159, Wafulz, Banedon, Cydebot, Evilal, Wfaxon, Doug Weller, Barticus88, Cetinsert, Archeus, Bugbear6502, Theqool, HappyInGeneral, Sean7phil, Parsiferon, Spud Gun, AntiVandalBot, Fayenatic london, Myanw, MurunB, Deflective, Scottcmu, Skomorokh, Bluerondo, Water bottle, Wasell, VoABot II, Careless hx, Tzq99, Violentbob, Jim Douglas, Skinnedmink, A1trips, SalvNaut, Malmckay, Dark archeus, Onecall, Kkett, Jamisonjo, MartinBot, Gandydancer, Fdimer, CommonsDelinker, PerpetuumMobile, Sebcastle, Jefebu, Jarry1250, HighKing, WLRoss, Squids and Chips, SimDarthMaul, VolkovBot, Christophenstein, JimmyHat, Aesopos, Zimbardo Cookie Experiment, Surrealmonk, Noformation, Rip969, AlleborgoBot, Theavex, CameraReady, SpitefulGOD, BenSeattle, Ctbrown101, Oxymoron83, Lightmouse, Maryyugo, Edavepitman, Frankj12, Superflush, Explanator, Sfan00 IMG, ClueBot, DFRussia, PaulLowrance, Amnw14545, Multipole~enwiki, Bbolingbroke, CohesionBot, Dekisugi, DumZiBoT, Benjaminbruheim, XLinkBot, Aopchuck, Kaiwhakahaere, Addbot, Donhoraldo, Johnhunsley, WikiUserPedia, LarryJeff, OlEnglish, Flukas, Cameron Scott, Subzerojones, Erik9bot, Legobot III, Aineolach, Biker Biker, Lars Washington, Irrito, Tideswimmer, RjwilmsiBot, Filmacu, EmausBot, Cjizzled, Stielzephyr, H3llBot, Sman9356, Saralicia, Marioluigi98, Helpful Pixie Bot, BG19bot, RobdotFran, Zedshort, Khazar2, Monkbot, Robert92107, Bizmaveric, Dubliner2015 and Anonymous: 299

19.6.2 Images

- **File:Ambox_current_red.svg** *Source:* https://upload.wikimedia.org/wikipedia/commons/9/98/Ambox_current_red.svg *License:* CC0 *Contributors:* self-made, inspired by Gnome globe current event.svg, using Information icon3.svg and Earth clip art.svg *Original artist:* Vipersnake151, penubag, Tkgd2007 (clock)

- **File:Ambox_important.svg** *Source:* https://upload.wikimedia.org/wikipedia/commons/b/b4/Ambox_important.svg *License:* Public domain *Contributors:* Own work, based off of Image:Ambox scales.svg *Original artist:* Dsmurat (talk · contribs)

- **File:BesslerOrffyreusWheel.png** *Source:* https://upload.wikimedia.org/wikipedia/en/9/9c/BesslerOrffyreusWheel.png *License:* Public domain *Contributors:* ? *Original artist:* ?

- **File:Boyle'{}sSelfFlowingFlask.png** *Source:* https://upload.wikimedia.org/wikipedia/commons/3/3b/Boyle%27sSelfFlowingFlask.png *License:* Public domain *Contributors:* ? *Original artist:* ?

- **File:Brillouin-Paradoxon.svg** *Source:* https://upload.wikimedia.org/wikipedia/commons/e/e8/Brillouin-Paradoxon.svg *License:* CC BY 3.0 *Contributors:* Own work *Original artist:* wdwd

- **File:Commons-logo.svg** *Source:* https://upload.wikimedia.org/wikipedia/en/4/4a/Commons-logo.svg *License:* ? *Contributors:* ? *Original artist:* ?

- **File:Cox_timepiece_winding_switch.png** *Source:* https://upload.wikimedia.org/wikipedia/commons/5/58/Cox_timepiece_winding_switch. png *License:* CC BY-SA 3.0 *Contributors:* Own work *Original artist:* J. .D. Redding

- **File:Crystal_energy.svg** *Source:* https://upload.wikimedia.org/wikipedia/commons/1/14/Crystal_energy.svg *License:* LGPL *Contributors:* Own work conversion of Image:Crystal_128_energy.png *Original artist:* Dhatfield

- **File:Edit-clear.svg** *Source:* https://upload.wikimedia.org/wikipedia/en/f/f2/Edit-clear.svg *License:* Public domain *Contributors:* The *Tango! Desktop Project*. *Original artist:*
  The people from the Tango! project. And according to the meta-data in the file, specifically: "Andreas Nilsson, and Jakub Steiner (although minimally)."

- **File:Feynman_ratchet.png** *Source:* https://upload.wikimedia.org/wikipedia/commons/e/ed/Feynman_ratchet.png *License:* Public domain *Contributors:* Own work *Original artist:* Bdkoivis

- **File:Flag_of_the_United_States.svg** *Source:* https://upload.wikimedia.org/wikipedia/en/a/a4/Flag_of_the_United_States.svg *License:* PD *Contributors:* ? *Original artist:* ?

- **File:Fulton.jpg** *Source:* https://upload.wikimedia.org/wikipedia/commons/f/f6/Fulton.jpg *License:* Public domain *Contributors:* ? *Original artist:* ?

- **File:Garabed_T._K._Giragossian_seated_reading.jpg** *Source:* https://upload.wikimedia.org/wikipedia/commons/f/f0/Garabed_T._K. _Giragossian_seated_reading.jpg *License:* Public domain *Contributors:* This image is available from the United States Library of Congress's Prints and Photographs division under the digital ID hec.09702.
  This tag does not indicate the copyright status of the attached work. A normal copyright tag is still required. See Commons:Licensing for more information. *Original artist:* Harris-Ewing collection

- **File:Johann_Bessler.gif** *Source:* https://upload.wikimedia.org/wikipedia/en/d/d2/Johann_Bessler.gif *License:* PD *Contributors:* ? *Original artist:* ?

- **File:Lightbulb.jpg** *Source:* https://upload.wikimedia.org/wikipedia/commons/2/29/Lightbulb.jpg *License:* CC-BY-SA-3.0 *Contributors:* Own work *Original artist:* Ming888 at English Wikipedia

- **File:MEGcircuit.png** *Source:* https://upload.wikimedia.org/wikipedia/commons/e/e0/MEGcircuit.png *License:* CC-BY-SA-3.0 *Contributors:* Transferred from en.wikipedia to Commons by Sfan00_IMG using CommonsHelper. *Original artist:* Reddi at English Wikipedia

19.6.3   Content license